UA Documents
Vol.1

Back to the Earth

2022. 12. 22 – 12. 27
K'ARTS Gallery

UA Documents
Vol.1

Harim Kang	강하림
Bogyong Kim	김보경
Seojin Kim	김서진
Jiseung Shin	신지승
Yonu Lee	이연우
Eunhoo Lee	이은후
Jiyun Lee	이지윤
Heesoo Jeon	전희수
Eunsol Jo	조은솔

차례

9	<u>파트 1.</u>	
10	서문	/ 우동선
18	기획의도	/ 김태영
22	읽기 전에	/ 김수영
		김태영
		김나리
		서승모
		캘빈츄아

35	<u>파트 2.</u>	
36	읽는 방법	/ 김태영
50	인덱스 카드	/ 이지윤

63	<u>파트 3.</u>	
64	읽는 방법	/ 이지윤
67	포트폴리오	/ 강하림
99		/ 김보경
131		/ 김서진
163		/ 신지승
195		/ 이연우
223		/ 이은후
253		/ 이지윤
285		/ 전희수
315		/ 조은솔

345	<u>부록</u>
346	전시기록
354	맺으며
356	감사한 분들
360	약력
367	크레딧

Contents

9	Part 1.	
10	Preface	/ Don-Son Woo
18	Introduction	/ Taeyoung Kim
22	Before You Read	/ Sooyoung Kim
		Taeyoung Kim
		Naree Kim
		Seungmo Seo
		Calvin Chua
35	Part 2.	
36	How to Read 1	/ Taeyoung Kim
50	Index Cards	/ Jiyun Lee
63	Part 3.	
64	How to Read 2	/ Jiyun Lee
67	Portfolio	/ Harim Kang
99		/ Bogyong Kim
131		/ Seojin Kim
163		/ Jiseung Shin
195		/ Yonu Lee
223		/ Eunhoo Lee
253		/ Jiyun Lee
285		/ Heesoo Jeon
315		/ Eunsol Jo
345	Appendix	
346	Exhibition Archives	
354	Closing	
356	Acknowledgments	
360	Biography	
367	Credit	

파트 1.
Part 1.

Sooyoung Kim Studio	Taeyoung Kim + Naree Kim Studio	Seungmo Seo Studio	Calvin Chua Studio	Themes, Values and Logic
◐				Architectural structure to predict and respond to changes throughout the entire life cycle of a building
	◐		◐	Environmental and social infrastructure leading to possibilities of architecture
◐	◐	◐		Potentials of adaptation and temporality leading to urban changes
		●	●	The locality to define our spatial and urban interventions

Table1. The Themes of the 5th-year Design Studios in 2022

김수영 스튜디오	김태영+김나리 스튜디오	서승모 스튜디오	캘빈츄아 스튜디오	주제 가치와 논리
◐				건축물의 전 생애주기 동안의 변화를 예측하고 대응하는 방식을 어떻게 건축적으로 구조화할 수 있는가
	◐		◐	환경, 사회 인프라스트럭처를 함께 생각함으로써, 건축은 어떤 가능성을 실현할 수 있는가
◐	◐	◐		사용자에 의한 변용 가능성/시간성이 건축의 의도를 지속하면서도 새로운 도시의 변화를 이끌 수 있는가
		●	●	우리의 도시·동네의 프로필과 캐릭터로부터 어바니즘을 끌어낼 수 있는가

표1. 2022년 건축과 5학년 설계 스튜디오 주제

The passage discusses the pursuit of good values and sound logic by architects in creating excellent architecture. It emphasizes that even though architects continually build logic and establish perspectives, identical values and logic do not guarantee identical architectural outcomes. The architect's sensibility and intuition come into play, adding a unique touch to each creation.

In a special feature on "Architect's Graduation Design" in the March 2020 issue of "Architecture" magazine, the author notes that graduation design provides a rare opportunity for architects to design freely, targeting a virtual client. It serves as a ritual to consolidate individual architectural knowledge and sensibilities developed during academic years, becoming the starting point for all future design work. Graduation design also acts as a litmus test to reaffirm one's aptitude and potential, a crossroads to gauge future career paths.

The architectural school is portrayed as a place where the values and logic of teachers are imparted to students. The passage expresses the intrigue in examining how the values and logic of teachers manifest in students' graduation projects, particularly within the context of the traditions and identity of K-Arts architecture The text concludes by mentioning the values and logic pursued by the 5th-year studio in the academic year 2022. However, the specific values and logic pursued in this studio are not provided in the provided text.

Table1. The themes of the 5th-year design studios in 2022

건축가는 좋은 건축을 하기 위하여 좋은 가치와 정연한 논리를 추구하기 마련이다. 건축가는 끊임없이 논리를 구축해 나가고 방법을 터득하며 관점을 수립해 나간다. 그렇지만 가치와 논리가 같다고 해서 곧 똑같은 건축물이 되는 것이 아니다. 그 사이에는 건축가의 감성과 감각이 개입하기 마련이다.

나는 '건축가의 졸업설계'를 특집으로 기획하면서 『건축』 2020년 3월호에 이렇게 썼다. "주지하듯이, 졸업설계는 가상의 건축주를 대상으로 하기에 건축가로서 모든 면에서 자유롭게 설계할 수 있는 매우 드문 기회일뿐더러, 학창 시절에 저마다의 건축지식과 감각을 집대성하는 통과의례로서 그 후의 모든 설계 작업의 원점이 된다. 또 졸업설계는 자신의 적성과 가능성을 새삼 확인하는 리트머스 시험지로서 혹은 시금석으로서 이후의 진로를 가늠하는 갈림길이 된다."

건축학교는 선생의 가치와 논리를 학생에게 전수하는 곳인데, 선생의 가치와 논리가 학생들의 졸업작품에 어떻게 드러나는지는 살피는 일은 흥미진진하지 않을 수가 없다. 여기에 한예종 건축과의 전통과 정체성이 놓이는 것만 같다. 2022학년도의 5학년 스튜디오는 다음과 같은 가치와 논리를 추구하였다.

표1.
2022년 건축과 5학년 설계 스튜디오 주제

In these statements, one can identify 14 keywords, including infrastructure, environment, society, aesthetics, users, transformation, temporality, architects, proposal, framework, intention, essence, city, and urbanism. These keywords seem to encapsulate the challenges of the 21st century, marked by capitalism and environmental issues due to climate change. These keywords are intricately linked to the concepts of place and type.

Place emerged as a counter to the spatial notions of modernism, while type emerged in opposition to the formal aspects in art history. Just with these concepts, it is evident that the 5th-year studio naturally aligns itself with the postmodern and post-20th-century era.

This semester, I am reading "The Seven Lamps of Architecture" (1849) written by John Ruskin (1819–1900) with students in the graduate program. The reason for delving into this book is that John Ruskin articulated concepts still prevalent in architecture today, 174 years ago. In this book, he evaluates seven lamps—sacrifice, truth, power, beauty, life, memory, and obedience. All of these values are Christian in nature, and as Ruskin highly values Gothic architectural forms, it may be challenging for me and the students to agree with every aspect of the book. However, we can find sharp insights in various corners.

In addition to the translated version, I looked for the English edition, and it's surprising that the English of this book, first published in 1849, is still relatively easy to read. This might be due to the minimal changes in the English language over time compared to the significant changes in Korean literature. Furthermore, after liberation, South Korea's education system leaned towards the Anglo-American model, contributing to this phenomenon.

이 말들에서 인프라, 환경, 사회, 미학, 사용자, 변용, 시간성, 건축가, 제안, 골격, 의도, 본질, 도시, 어바니즘과 같은 14개의 키워드들을 골라낼 수가 있다. 이 키워드들에는 지금 우리가 처한 21세기의 자본주의와 기후 변화로 인한 환경 문제가 들어 있는 것만 같다. 그리고 이 키워드는, 다시 장소와 유형이라는 개념과 결부한다.

장소는 주지하다시피 모더니즘의 공간에 대항하여 등장하였고, 유형은 예술사의 양식에 대항하여 등장하였다. 이 개념들만으로도 5학년 스튜디오가 당연하게도 포스트모더니즘 이후와 20세기 이후를 지향하고 있음을 알 수가 있다.

나는 이번 학기에 예술전문사 과정의 학생들과 존 러스킨(John Ruskin, 1819–1900)이 30세에 쓴 『건축의 일곱 등불 The Seven Lamps of Architecture』(1849)를 강독하고 있다. 이 책을 읽는 이유는 존 러스킨이 지금도 통용하고 있는 건축의 개념을 174년 전에 기술해 놓았기 때문이다. 이 책에서 그는 희생, 진실, 힘, 아름다움, 생명, 기억, 복종이라는 등불을 각각 높이 평가하고 있다. 이 일곱 가치는 모두 기독교적인 것들이며 그가 높이 평가하는 건축 양식은 고딕이라서, 나와 학생들은 이 책의 모든 내용에 동의하기가 어렵지만 구석구석에서 날카로운 통찰을 만날 수가 있다.

번역판에 더하여 영어판을 찾아보는데, 1849년에 처음 출간된 이 책의 영어를 지금도 그리 어렵지 않게 읽을 수가 있다는 것이 새삼 놀랍다. 이는 그간에 영어는 변화가 적었던 것에 반하여 한국의 서적은 변화가 극심하였고, 해방 후 한국의 교육이 영미권을 지향하고 있었기 때문에 생기는 일일 것이다.

Meanwhile, my late mentor Suzuki Hiroyuki (鈴木博之, 1945–2014) wrote "The Seven Powers of Architecture" (1984) at the age of 39, consciously echoing the title of John Ruskin's "The Seven Lamps of Architecture." Suzuki Hiroyuki's seven powers are Imagination, Numbers, Gothic, Detail, Imitation, Command, and Tradition. These values are not only the principles that create architecture but also the principles that describe architecture.

In his work "How to Look at Modern Architecture" (1999), written at the age of 54, Suzuki Hiroyuki discussed 12 values: Decoration, Function, Machinery, History, Structure, Architect, Nature, Moral, Art, Space, Map, and Place. These values allow architects to demonstrate their uniqueness while also finding common ground. Regarding the concept of 'Place,' which overlaps with the theme of graduation projects at K-Arts, Suzuki wrote, "Architecture is never a universal presence. Architecture gains meaning only when it contributes to the place where it stands."

Sacrifice, Truth, Beauty, Life, Memory, Obedience, Imagination, Numbers, Gothic, Detail, Imitation, Command, Past, Decoration, Function, Machinery, History, Structure, Architect, Nature, Moral, Art, Space, Map, Place-these 26 values individually contribute to creating good architecture and form the foundation of our contemporary architectural perspective.

(When I was commissioned to write the manuscript, I intended to discuss the correlation between the previous 14 keywords and these 26 values, but due to limitations in time and space, I couldn't pursue that.)

한편 내 은사인 스즈키 히로유키(鈴木博之, 1945-2014)가 39세에 쓴 『건축의 일곱 가지 힘 建築の七つの力』(1984)은 그 제목에서 존 러스킨의 『건축의 일곱 등불』을 의식하였다. 스즈키 히로유키가 말하는 일곱 가지 힘이란 연상, 수, 고딕, 세부, 모방, 지령, 과거이다. 이 가치들은 건축을 만드는 가치이기도 하지만, 건축을 설명하는 가치이기도 하다.

스즈키 히로유키가 54세에 쓴 『현대건축을 보는 법 現代建築の見かた』(1999)에서는 12가지의 가치를 논하였다. 장식, 기능, 기계, 역사, 구조, 건축가, 자연, 모럴, 아트, 공간, 지도, 장소. 이러한 가치들로 건축가들의 고유성이 발휘되며, 그 점에서 또 공통성이 있다고 보았다. 한예종 졸업작품의 주제와 겹치는 개념인 '장소'에 대해서, 스즈키는 "건축은 결코 보편적인 존재가 아니다. 건축은 그것이 서는 장소에 기여하여 비로소 의미를 갖는 것이다"라고 썼다.

희생, 진실, 아름다움, 생명, 기억, 복종, 연상, 수, 고딕, 세부, 모방, 지령, 과거, 장식, 기능, 기계, 역사, 구조, 건축가, 자연, 모럴, 아트, 공간, 지도, 장소. 이 26개의 가치는 각각 좋은 건축을 이루었고 현대의 우리가 갖는 건축관의 배경이 되고 있다.

(원고를 의뢰받았을 때는, 앞의 14개의 키워드와 이 26개의 가치의 상관관계를 논하려고 마음먹었었는데, 시간과 지면이 부족하여 그러지 못하였다.)

Harim Kang belongs to Kevin Chua Studio and created a work titled "Cul-de-sac_ Neutral Space as a Typology of Knowledge Sharing Lab." She defined neutral space as a typology of knowledge sharing lab within the city and developed this organization. The core focus of the studio is to reconsider the approach to the 15-minute city from the perspective of strategic intensification of Seoul's existing areas through the concept of 'urban rooms.' Within the existing scales of micro and macro in the urban rooms, she proposes a new 'urban room' that benefits 1000 local residents.

Bogyong Kim is part of Seungmo Seo Studio and submitted a work titled "Magnetic Field City." She responded to the studio's direction by exploring ways to enjoy the city and architecture from the user's perspective and incorporating 'play' for new architectural plans. To alleviate the rigidity of the city and discover new urban planning with dynamic flows, she adopted Sudoku. The process of finding numbers in the blank spaces drew attention to the repetitive nature of temporary flows across the axes of the Sudoku board. The temporary flow of Sudoku was substituted for moments of temporary solidarity in the city. Thus, she explored ways for wedding ceremonies to intervene in the city as a program for expressing temporary solidarity in urban spaces.

Seojin Kim belongs to Taeyoung Kim + Naree Kim Studio. "Rain Festival" is a reflection on a life accompanied by plants. The story began with the experience of introducing plants into a small room, and the expectation was for this experience to expand into the city. The increasing social inequality due to climate crises revealed vulnerability to rain, and the project envisions regions with vulnerability to rain overcoming these challenges and forming communities that care for each other.

강하림은 캘빈츄아 반에 속하며, "Cul-de-sac_Neutral Space as a Typology of Knowledge Sharing Lab"이라는 제목의 작품을 만들었다. 그는, "도시의 한 유형으로 중립적 공간(neutral space)를 정의하였고, 이러한 조직을 추출한다. 추출한 중립적 공간을 건축화한다. 스튜디오의 핵심 관심사는 '도시의 방'을 통한 서울 기존 지역의 전략적 강화의 관점에서 15분 도시에 대한 접근 방식을 재고하는 것이다. 기존의 도시의 방이 가지는 미시와 거시의 두 스케일 사이에서, 1000명의 지역주민에게 혜택을 주는 새로운 '도시의 방'을 제안한다."

김보경은 서승모 반에 속하며, "Magnetic Field City"라는 작품을 제출하였다. 그는, "사용자의 관점에서 도시, 건축을 즐기는 방식을 찾고 새로운 건축계획을 위해 '놀이'를 차용하여 도시 리서치를 시작하는 것이 스튜디오의 방향이었다. 도시의 경직성을 해소하고 새로운 흐름을 가진 도시계획을 찾기 위해 스도쿠를 채택하였다. 빈칸의 숫자를 찾아 나가는 과정은 스도쿠판 축의 경계를 가로지르는 일시적인 흐름의 반복이라는 점에 주목하였다. 스도쿠의 일시적인 흐름은 도시에서 일시적인 유대감을 느끼는 순간으로 치환하였다." 그리하여 "일시적 유대감을 발현할 도시공간의 프로그램으로 결혼식이 도시에 관입할 방식을 모색하였다."

김서진은 김태영+김나리 반에 속한다. "Rain Festival"은 식물과 함께하는 삶에 대한 고찰에서 시작된 이야기이다. 작은 방에 식물을 처음 들였을 때의 경험이 도시로 확대되는 것을 기대했다. 기후 위기로 인한 불균등은 곧 사회적 불평등으로 드러났고, 비에 대한 취약함을 가진 지역이 그것을 극복하고 서로를 돌보는 공동체가 되는 바람을 담았다.

Jiseung Shin is part of Taeyoung Kim + Naree Kim Studio, and she created "The Mounds, Artifacts." The emphasis in her work is on the concept of 'ground,' which was significant in the project, and it is highlighted on the horizontal plane of the exhibition furniture. The furniture has a long horizontal surface of 2500×6000, providing an extended space for exhibition. The design aims to encourage viewers to face the wall directly, ensuring a prolonged and immersive experience rather than passing by casually.

Yonu Lee belongs to Taeyoung Kim + Naree Kim Studio, and she created "Dongmyo: Intergenerational Network." She expressed a long-standing interest in boundaries since her school days. Exploring Dongmyo Market, she discovered the potential for breaking down generational barriers in an aging society. This realization became the driving force behind her work. Reflecting on her words, it's interesting to note that Yonu Lee had previously written a paper on the elderly during his studies. The symbol '\-' in the title symbolizes both the disruption and connection of boundaries.

Eunhoo Lee is part of Sooyoung Kim Studio, and she submitted "The Fourth Attack: Linking Instabilities." The project aims to connect elements threatening Yongsan, a land characterized by instability due to water and development. It envisions a response to crisis directly manifested in the city, offering an imaginative solution to alleviate the anxiety about the impermanence of the land. The process parallels how human cognition leads to preparation.

신지승은 김태영+김나리 반에 속하며, "The Mounds, Artifacts"를 제작하였다. 그는, "전시하는 것 자체가 작업의 연장이다. 작업에서 중요했던 화두 '땅'이, 전시 가구에서도 강조돼 보이는 수평면에서 드러나도록 했다. 2500×6000의 가로로 긴 자리였다. 6000만큼 벽에 붙어있는 자리였기 때문에, 관람객이 지나가다 흘겨보지 않고, 벽을 정면으로 마주하게 하며, 오래 머물게 하는 것이 목표였다."

이연우는 김태영+김나리 반에 속하며, "Dongmyo: Intergen\-erational Network"를 만들었다. 그는 "학창시절 내내 경계에 관해 관심이 많았"는데, 동묘시장에서 초고령사회에서 "세대간의 경계가 허물어질 수 있는 가능성을 보"았다고 했다. 그것이 이 작업의 추동력이라고 하였다. 이 말을 읽고 돌이켜 보니까, 이연우는 내 반에서 노인들을 주제로 논문을 작성했었다. '\-'은 경계의 단절과 연결을 기호화한 것이다.

이은후는 김수영 반으로, "The Fourth Attack: Linking Instabilities"를 제출하였다. "본 프로젝트는 물과 개발로부터 불안정한 용산이라는 땅에 위협이 되는 요소를 하나로 잇는 방향으로 흐른다. 위기와 대응이 도시에 그대로 드러나서, 땅이 영원하지 않을 것이라는 불안이 해소될 방법이 되는 상상을 한다. 인간의 인지가 준비로 이어지는 것과 같다."

Jiyun Lee belongs to Seungmo Seo's studio, and she argued for "Plowing and Blowing." As the title suggests, she brought up "Plowing" as a stepping stone to his deep interest in the land and the infinite possibilities of student works. Blowing is derived from the Chinese word for "wind". The strategy she chose was, "In the process of digging the underground layer, a tunnel method is used to create a large underground space by first building a tunnel in the lower part, and the upper structure is a stacked dome, and the soil dug from the underground is stacked between the dome structures to create an uneven soil core to secure the diversity of vegetation." Jiyun Lee was in my class and wrote a paper on the domes of Greek Orthodox churches in Korea. Her theoretical exploration of domes led to her graduation design.

Heesoo Jeon is part of Taeyoung Kim + Naree Kim Studio, and she created "Tip-Toeing Landscape." With a deep interest in Brutalism, she has previously written a paper on Brutalism in the United Kingdom. Thus, "This project imagines a new urban form capturing the essence of the current era on the cusp of a new age. In a land marked by the inequalities created by capitalism, urban dwellers initiate small rebellions. Defying gravity, tiptoeing with heels, they stride toward a new world." The project stems from a study initiated with megastructures, seeking solutions for polarized cities and proposing a new urban paradigm that may emerge in the future. Though at times bold and seemingly unrealistic, the project aims to become a driving force to stir stagnant humanity in various ways.

이지윤은 서승모 반으로, "Plowing and Blowing"을 주장하였다. 제목에서 드러나듯이, 그는 대지에 대한 깊은 관심과 학생작품의 무한한 가능성을 디딤돌로 삼아 "농법(Plowing)"을 들고나왔다. 블로잉(Blowing)은 "풍백(風伯)"에서 따온 것이라고 읽힌다. 그가 택한 전략은, "지하층을 파내는 과정에서는 하부에 갱도를 먼저 조성하는 터널 공법을 통해 지하 대공간을 조성하고, 상부 구조는 돔이 적층되는 형태로 돔 구조의 사이에 지하에서 파낸 흙을 쌓아 일정하지 않은 토심을 조성하여 식생의 다양성을 확보하는" 것이었다. 이지윤은 내 반에서 우리나라 그리스 정교 성당의 돔에 관한 논문을 썼었다. 돔에 대한 이론적 탐구가 졸업설계로 이어진 셈이다.

전희수는 김태영+김나리 반으로, "Tip-Toeing Landscape"를 만들었다. 그는 브루탈리즘에 깊은 관심을 가졌다. 내 반에서는 영국의 브루탈리즘에 관하여 논문을 작성했었다. 그리하여 "이 프로젝트는 새로운 시대의 과도기에 선 지금, 이 시대의 모습을 담은 새로운 양식의 도시를 상상으로 그려냈다. 자본주의가 만들어낸 불평등한 땅에서 도시인들은 작은 반항을 시작한다. 중력에 맞서 뒤꿈치를 쳐든 채, 까치발로 새로운 세상을 향해간다." 다음과 같은 당부가 있었다. "이 작품에서는 메가스트럭쳐로부터 시작한 스터디에서 양극화된 도시에 대한 해답을 찾고, 미래에 등장하게 될 새로운 도시 패러다임을 제시한다. 다소 과감하고, 때로는 비 현실적으로 비춰질 수 있으나, 여러 방면에서 새로울 이 시도가 정체된 인간을 다시금 움직이게 할 동력이 되길 바란다."

Eunsol Jo, a member of Taeyoung Kim + Naree Kim Studio, submitted a work titled "Dance Rock-et." The project proposes a new form of station in the triangular area with long unused transfer passages in Samgakji. It harnesses energy from subway vibrations and people's footsteps in this area designated solely for transfers. The sense of travel from vibrations and wind creates a unique experience, turning it into a source of energy and a new form of production and labor. Analyzing the flow of air, vibrations, and sound and transforming them into spatial experiences, the project envisions an active public history created by individuals coming together, providing opportunities for fair and universal imagination. Targeting subway stations traditionally designed for functional purposes, the project assigns them previously overlooked roles of environmental, spatial, and experiential significance, proposing a new role for public spaces in Seoul.

The selected passage emphasizes the anticipated manifestation of the logic and values embedded in the 9 graduation projects. To understand how these principles are realized in practice, readers are encouraged to closely examine the drawings featured in this book. Furthermore, it's suggested to pay attention to the connections between these logics and values and those presented by the 4 studios.

Just last week, I had a conversation with someone about the phrase 'Emerging Excellence (青出於藍).' This ancient Eastern virtue seems to be a keyword explaining the development of present-day South Korea. Confirming the 'Emerging Excellence' of these 9 graduates could be a source of great joy in the future. Perhaps, the hue of 'Emerging Excellence' is already infused within the pages of this book.

조은솔은 김태영+김나리 반인데, "Dance Rock-et"이라는 작품을 제출하였다. "본 프로젝트는 환승의 용도로만 이루어져 긴 유휴 환승통로를 가진 삼각지역에 지하철의 진동과 사람들의 발걸음으로 에너지를 만들고 이를 경험하는 새로운 형식의 역사를 제안한다. 진동과 바람이 주는 여행의 감각은 에너지를 만들며 새로운 생산, 노동이 된다. 공기, 진동, 소리의 흐름을 분석하여 이를 공간의 경험으로 전환시키고 개개인이 모여 만들어내는 능동적인 공공역사는 공평하고 보편적인 상상의 기회를 제공한다. 기능적인 목적으로만 설계되어 온 지하철역을 대상으로, 기존에 고려되지 않았던 환경적 공간 경험적 역할을 부여하여 서울에서 제시해야 하는 새로운 공공공간의 역할을 제안한다."

이상에서 9명의 졸업작품의 설명에서 그들의 논리와 가치에 주목하여 발췌·인용하였다. 학생들의 글에 담긴 논리와 가치가 자현(自現)하기를 기대하였다. 독자들께서 이 논리와 가치가 실제 작품에서 어떻게 구현되었는지를 살피기 위해서는 이 책에 수록한 도면들을 더욱 들여다보아야 할 것이다. 그리고 그 논리와 가치들이 4개의 스튜디오가 제시한 논리와 가치와 어떻게 연결하는지를 주목하기를 바란다.

바로 지난주에 나는 어느 분과 '청출어람(青出於藍)'을 말한 적이 있다. 이 오래된 동양의 미덕이 지금 대한민국의 발전상을 설명해주는 하나의 키워드가 될 것 같다. 이 9명의 졸업생들의 '청출어람'을 확인하는 일을 미래의 커다란 즐거움으로 남겨두고 싶다. 어쩌면 이 책 안에 그 남색(藍色)이 이미 배어 있을지도 모르겠지만.

2023. 11. 1.

기획의도

김태영

The portfolio presents both the process and outcomes of exploring personal interests related to the studio theme, as well as the research and analysis required to develop these into a cohesive, original topic. This journey includes testing ideas across various media and culminates in a compelling final plan. While showcasing individual work, the portfolio also aligns with the studio's identity and the department's broader goals, reinforced through tutorials, critiques, workshops, and peer reviews that incorporate external insights and feedback.

In 2022, four advanced studios—Sooyoung Kim Studio, Taeyoung Kim + Naree Kim Studio, Seungmo Seo Studio, and Calvin Chua Studio—produced nine portfolios examining spatial concepts through research, reflection, and critique, pushing the boundaries of architectural skill. Their aim is to foster architects committed to environmental and social responsibility. Each studio's tutors shape its character and themes by presenting case studies, exploring relevant literature, and posing critical questions. The "UA Documents" platform allows for exploration of these nine portfolios in reverse order, showcasing the unique work of nine students and facilitating understanding and comparison.

포트폴리오는, 스튜디오 주제에 부합하는 개인의 관심사를 찾아내어 리서치와 분석을 통해 맥락화하여 독창성있는 주제로 발전시키고 이에 대한 전제를 세워 다양한 매체를 활용하여 테스트하며, 결국에는 설득력있는 계획으로 완성하는 과정과 결과물을 모두 포함한다. 각 포트폴리오는 개인의 고유한 작업이지만, 소속된 스튜디오 튜터와의 튜토리얼을 통한 발전, 학기별 중간 및 기말심사, 전문가 워크샵 및 피어리뷰를 거치면서 피드백의 선택 및 흡수를 거친, 스튜디오 정체성과 학과의 정체성 모두에 대응하는 다면적 작업이기도 하다.

2022년 4, 5학년을 묶어 버티컬 스튜디오로 운영하는 고급 스튜디오는 김수영 스튜디오, 김태영+김나리 스튜디오, 서승모 스튜디오, 캘빈츄아 스튜디오로 총 네 개의 스튜디오이다. 지적 호기심과 리서치, 성찰과 비평, 새로움을 추구하는 예술가의 시각으로 공간을 탐구하고, 건축 기능을 실천하는 방식을 창의적으로 확장하며, 환경과 사회에 대한 책임과 소통에 적극적인 건축가를 양성하는 학과 목표 아래 각 튜터는 지향하는 건축의 방향, 접근 방식에 따라 스튜디오의 성격과 우선 주제를 설정한다. 함께 참고사례를 연구하고, 관련 문헌을 읽으며, 스스로에게, 그리고 이 도시와 사회에서 유효한 질문을 찾는다.

이 책은 그 질문과 답을 찾아가는 여정으로서의 졸업생 9인의 포트폴리오를 담는다. 포트폴리오를 열고 넘기며 멈추고 들여다보는 공간적, 시간적 행위로 재생하며 포트폴리오를 회고하고 성찰하는 공간과 시간으로서 이 책을 기획한다.

기획의도 김태영

The starting point for understanding the similarities and differences among the nine distinct working methods and outcomes lies in three key criteria: keywords, places, and purposes. As certain works intersect, diverge, and reassemble, various collaborative reading methods emerge, uncovering connections that might otherwise go unnoticed. This reading process transcends the individual studios and reveals shared values that extend beyond the boundaries of each portfolio.

Studio tutors review the completed works, revisiting questions posed—whether implicitly or explicitly—in the studio briefs. Through their reflections, we engage in an active reading of each work, as if contemplating the brief prior to examining the portfolios. Across these nine portfolios, we explore the potential to enhance spaces, environments, cities, and societies. By addressing contemporary issues, we envision alternative possibilities and affirm the value of the freedom to imagine them. We invite you to navigate these possibilities, underscoring the importance of envisioning the feasibility of our dreams and understanding what we can achieve together.

9인의 서로 다른 작업방식과 결과물 간의 유사성, 차이를 이해하는 세 개의 기준으로서 키워드, 장소, 용도로부터 시작한다. 몇 개의 작업들이 서로 연결되고 해체되며 다시 연결되는 가운데 함께 읽기의 다양한 방식을 제안한다. 스튜디오를 넘어서 개인이, 우리가 중요하게 생각했던 가치들이 연결되는 지점이다.

우리는 9인의 포트폴리오를 통해 더 좋은 공간, 더 좋은 환경, 더 좋은 도시, 그리고 더 좋은 사회의 가능성을 탐색한다. 현재의 문제를 직시하고 도전하며 더 좋은 방식을 꿈꾸는 자유가 얼마나 가치있는가를 이야기하며 그 이야기에 초대한다. 우리의 꿈이 얼마나 가능한지, 우리가 무엇을 실천할 수 있는지를 조망할 수 있을 때, 우리의 미래는 더 밝을 수 있다.

How can the architectural structure be designed to predict and respond to changes throughout the entire life cycle of a building?

The current research on 'sustainability' exerts a significant influence across various fields, particularly in architecture. However, the predominant focus in contemporary architecture revolves around the internal aspects of programs, planning, and operations, with a primary emphasis on energy-saving awareness.

The core essence of architecture lies in proposing a structural framework that can adeptly adapt to both tangible and intangible changes occurring throughout a building's entire life cycle. To truly attain sustainability, a building must possess 'resilience' to withstand external shock.

건축물의 전 생애주기 동안의 변화를 예측하고 대응하는 방식을 어떻게 건축적으로 구조화할 수 있는가?

우리는 설계하는 데 있어 결과가 명백한 목표를 직접적으로 드러내는 것은 아니지만, 해석의 여지들을 가질 수 있기 때문에, 사용성을 통해 그 정체성이 드러날 수 있도록 해야 한다. 우리가 만들어야 하는 것은 하나의 제안이며, 이는 몇 번이고 특정한 상황에 적합한 구체적인 반응을 이끌어낼 수 있어야 한다. 그래서 이 제안은 단순히 중성적이고 포용적이어서는 안되며, 그렇다고 특별하지 않아서도 안 된다. 이것은 우리가 다기능이라고 부르는 더 넓은 효용성을 가져야만 한다.

현재 많은 영역에서 '지속가능성'에 대한 탐구들이 이루어지고 있고, 건축에서도 간과하기 어려운 중요한 영향력을 갖게 되었다. 그러나 현재 건축에서 활발하게 다루어지고 있는 것은 본질적인 영역이라기 보다 프로그램, 기획 혹은 운영에 대한 인테리어적인 성격이 주를 이루고 있거나 혹은 에너지 절약에 대한 경직된 인식이 주를 이루고 있다. 건축이 갖는 본질적 속성이란 건축물의 전 생애주기 동안에 일어나는 유, 무형의 변화에 대해 적절하게 대응할 수 있는 구조적인 제안이다. 결국 지속 가능한 건축물이 되기 위해서는 외부의 충격에 대해 '회복탄력성'을 갖추고 있어야 한다.

Given the dynamic and ever-changing conditions, accurately quantifying the life cycle of a building becomes a challenging task. It's essential to recognize that a prolonged life cycle may not always be the most suitable condition for sustainability. Nevertheless, 'resilience' plays a crucial role, especially for buildings with a relatively long life and high cost, unlike rapidly evolving industrial products that frequently necessitate replacement. Resilience imposes an obligation for analysis, prediction, and strategic application not only during the architectural planning and construction phases but also in the medium and short-term usage of the building. Various sustainable elements may be incorporated in some cases, while in others, they might be overlooked.

Architects need to prioritize the resilience of a building. By actively considering and predicting resilience, the overall value of the building can be significantly enhanced.

급변하고 있는 유, 무형의 조건들 속에서 건축물의 생애주기를 정확하게 정량화하는 것은 불가능할 뿐만 아니라 반드시 긴 생애주기가 지속가능성에 부합하는 조건이라고 생각하지는 않는다. 다만 '회복탄력성'은 지속적으로 진화하며 수시로 교체되어야 하는 공산품과 달리 비교적 긴 시간과 많은 비용을 필요로 하는 건축물이 반드시 감당해야 하는 역할이다. 그것은 건축가가 건축물을 계획하고 지어지는 영역뿐만 아니라 건축물이 사용되는 동안 중, 단기적으로 처하게 될 상황들을 분석하고 예측하여 전략적으로 건축물에 적용해야 할 의무를 갖는다. 다양한 지속가능한 요소들이 덧붙여지는 경우도 있지만 그 반대의 경우도 존재한다.

건축가는 건축물의 '회복탄력성'을 고려하여야 하고, 예측하려고 애를 쓸 때에 건축물에 대한 가치는 좀 더 확장될 것이라고 생각한다.

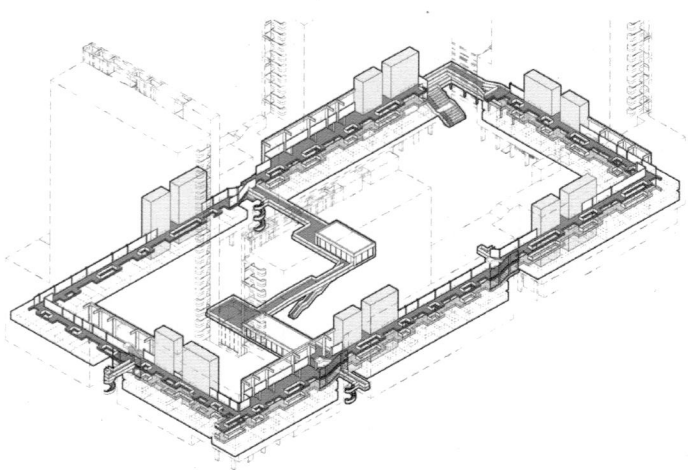

ⓒ 한 공공주택 설계공모 당선안에서 중정을 둘러싸며 반층의 차이로 분절되는 주거 클러스터와 이를 걸어 오르내리며 연결하는 복도의 순환으로 이웃의 규모와 연결성을 정의하게 하는 복도 공동체를 제안했다.

ⓒ U.topo The winning proposal for the public housing design competition of Wangsuk A-23 Block suggests a corridor community that defines the scale and connectivity of neighbors. It features residential clusters segmented by a half-level difference surrounding a courtyard and connected by circulating corridors that traverse up and down.

By considering environmental and social infrastructure together, what possibilities can architecture realise?

History reveals our propensity to rely on technology as a solution to problems, often resulting in the emergence of a new humanity shaped by that technology. In today's world, there is a growing belief that skilled hands working with tools and eyes that constantly adjust and mediate outcomes can protect humanity from various threats.

In a society where the visible guides us more than anything else, shaping our experiences and informing our judgments, what role does architecture play beyond mere visibility? What value can it offer? How and where can we create spaces that provide an authentic experience of the real world and enable us to engage with reality? Furthermore, how can we develop egalitarian and universal architecture that resonates with the beauty of space and shares its inherent values? The quest for answers to these questions involves proposing architecture as an infrastructure aligned with the following principles.

Human beings inherently desire exchange and empathy, as much as they strive for concentration and immersion. While people may compete for scarce resources or opportunities, they often achieve their aims through cooperation and collaboration. The feelings, analyses, judgments, decisions, experiences, and aspirations we navigate in this process are synesthetic. Architecture, in this context, embodies synesthesia, transcending what is merely seen or displayed. This form of architecture, which can be universally appreciated, encompasses a creative essence, environmental quality, and social value.

환경, 사회 인프라스트럭쳐(기반시설)를 함께 생각함으로써, 건축은 어떤 가능성을 실현할 수 있는가?

우리의 역사는, 문제에 대한 해결책으로 기술에 기대고, 그 기술의 결과 새로운 인류를 만들어 온 과정에 익숙함을 보여준다. 특히 이 시대에는, 도구를 놀리는 정교한 손과, 그 결과물을 끊임없이 조정하고 중재하는 눈이야말로 위협으로부터 인류를 지켜낼 수 있다는 믿음이 더욱 깊어지고 있다.

다른 어떤 것보다 보이는 것이 안내하는 세상, 보여지는 것으로 경험하는 세상, 볼 수 있어야 판단하는 세상에서 그 너머의 건축은 무엇이며, 어떤 가치를 가질 수 있을까? 실제 세계의 경험을 주는 공간, 실재성을 경험하게 하는 공간은 어떻게 어디에 만들어져야 하는가. 그 공간의 아름다움을 공감하고, 그 가치를 공유하게 하는, 평등하고 보편적인 건축은 어떻게 가능한가. 그에 대한 질문과 대안을 찾아가는 과정은 다음의 전제와 그에 대응하는 인프라스트럭쳐로서 건축을 제안하는 것이다.

인간은 집중과 몰입을 추구하는 만큼 교류와 공감을 바란다. 부족한 자원이나 기회를 향해 경쟁하지만 협조와 협력을 병행한다. 그 과정에서 거치는 느낌과 분석, 판단과 결정, 경험과 전망은 공감각적이다. 공감각적 건축은 보이는 것 또는 보여지는 것 너머의 건축을 지시한다. 그리고 보편적으로 향유할 수 있는 공감각적 건축은 창의적 감각, 환경의 퀄리티, 사회적 가치를 가진다.

ⓒ 유블로 자재를 최소화해 간결하게 만든 유블로 창호 시스템. 고층 아파트에 보편화된 복잡한 난간 분할창 대신, 추락이 불가능한 작은 환기창이 있는 통창으로 간결하게 리모델링했다. 다양한 재질과 기능의 매개는 개인의 취향을 담아 선택할 수 있다.

ⓒ UBLOUBLO window system, designed to simplify the component to minimize the use of frames, and to create energy efficient ventilation system. In stead of complex divided windows commonly found in high-rise apartments, they were remodeled with simple ventilation windows that prevent falls. Various materials and functional window covers can be selected according to personal preferences.

Can we discuss architectural aesthetics while dealing with the environment and society?

The facade is the building's attire. Like fashion, the facade serves as a means of expressing aesthetics while also protecting and regulating the interior from the external environment. Traditionally, in Western architecture, the facade as the 'face' of the building, emphasized aesthetics such as symmetry and proportion. However, today's facade must function as an environmentally and socially inclusive exterior that encompasses materials and technologies beyond mere appearance.

So, can we discuss architectural aesthetics while dealing with the environment and society? Just as Francis Kéré stated, it is not because you are rich that you should waste material. It is not because you are poor that you should not try to create quality. Everyone deserves quality, everyone deserves luxury, and everyone deserves comfort. Socially and environmentally conscious architecture can undoubtedly be beautiful. Contemporary architectural aesthetics merely differ from the aesthetics of the past, fitting the spirit of our times.

The facade, which delineates the boundary between the interior and exterior, comprises the external facade and the internal facade. Most buildings, like traditional Western architectural facades, are designed with a primary focus on the external facade, with the internal facade being a by-product of the external facade. In contrast, in traditional Korean architecture, I believe there was a greater emphasis on the internal facade, where people in the space were connected to the exterior through windows.

As the need for more internal environmental control such as shading, lighting, ventilation, and insulation increases, the significance of the internal facade becomes more pronounced, as was the case in traditional Korean architecture. Today, the facades that reflect the world we live in seem to prioritize the interior over the exterior, the materials and construction methods over the finishes, simplicity over complexity, and the preference for aged but valuable items over cheap, disposable ones. Naturally, this aesthetics will become our new aesthetics.

환경과 사회를 다루면서 건축 미학을 논할 수 있는가?

파사드는 건물의 옷이다. 패션처럼 파사드도 미학을 표현하는 수단이자 외부 환경으로부터 내부를 보호하고 조절하는 역할을 한다. 전통적으로 서양 건축의 '얼굴'로서의 파사드는 대칭, 비례와 같은 미학을 중시했지만 오늘날 파사드는 그 어떤 때 보다 보여지는 것 너머의 재료와 기술을 아우르는 사회 환경적 외피로 기능해야 한다.

그렇다면 환경과 사회를 다루면서 건축 미학을 논할 수 있는가? 프란시스 케레가 이야기했듯 부자라고 자원을 낭비할 권리가 없는 것처럼, 가난해도 귀하고 안락한 것을 누릴 권리가 있다. 사회 환경적 건축도 당연히 아름다울 수 있다. 단지 지금 시대 정신에 맞는 건축 미학은 예전의 미학과 달라질 뿐이다.

내외부의 경계를 만드는 파사드는 외부 파사드와 내부 파사드를 형성한다. 대다수의 건축물은 전통적인 서양 건축 파사드처럼, 외부 파사드 위주로 설계되고 내부 파사드는 외부 파사드의 부산물이다. 반면 한국 전통 건축에서는 공간 속 사람이 외부와 창으로 연결되는 내부 파사드를 더 우선시했다고 생각한다.

차양, 채광, 환기, 단열 등 점점 더 많은 내부 환경 제어가 필요할수록 한국 전통 건축에서 그랬던 것처럼 내부 파사드가 중요해진다. 오늘 우리가 사는 세상을 반영한 파사드는 외부보다 내부를, 마감보다 그 재료나 공법을, 복잡함 보다 간결함을, 싸고 쉽게 버려지는 것 보다 낡지만 귀한 것을 지향해야 하지 않을까? 그리고 자연스럽게 그 미학은 우리의 새로운 미학이 될 것이다.

In the rapidly changing landscape of society, where user autonomy and adaptability are on the rise, can architects' proposals (intent, essence, framework) be sustainable?

1. When Architecture is Needed

In La Fontaine's Dogville, a film that is also theatrical, the building plan is drawn on the floor in a 1:1 scale. Inside this framework, the actors perform as if the building itself is present. The building plan is abstract yet concrete because human life continues within this framework. The reason we engage in architecture begins with adding appropriateness to the urban context or eliminating unnecessary spaces. The desired outcome is a comfortable and non-monotonous structure. It is akin to a daily meal, akin to a formal meal centered around rice. Unassuming and not provocative, providing comfort to the body, allowing us to taste the natural essence. If architecture is like a meal, life is like a side dish. I anticipate a life in the city, much like the diverse array of flavors that side dishes create in a formal meal. Currently, architecture, like a formal meal that cannot be complete without rice, is needed. People need a place to reside.

2. Attitude Becomes Form

"It's often said, 'Attitude becomes form.' The legendary curator Harald Szeemann's 1969 exhibition 'When Attitudes Became Form' at the Kunsthalle Bern is a historic moment that not only changed the language of exhibitions but also the course of contemporary art history. While discussing architecture, the city, and neighborhoods, the word 'attitude' might seem somewhat out of place."

This is a term chosen by my longtime friend, Hong Borah, during our conversation about neighborhoods. She quoted a sentence that she used to describe me.

시시각각 급변하는 사회, 그리고, 사용자의 자율성, 변용가능성이 커가는 지금, 건축가의 제안(의도, 본질, 골격)은 지속가능한가?

1. 건축이 필요할 때
라스폰트리에의 도그빌은 영화이면서 연극적이다. 건물 평면을 1:1로 바닥에 그려 놓고, 그 안에서 배우들은 건물이 있는 듯 연기한다. 건물 평면은 추상적인 동시에 구상적이다. 왜냐하면, 이 틀을 배경으로 인간의 삶은 지속되기 때문이다. 우리가 건축을 하는 이유는 도시의 적재적소에 적절함을 더하거나, 불필요한 곳을 덜어내는 것으로부터 시작한다. 그렇게 나온 결과물이 편안하고, 지루해보이지 않기를 바란다. 마치 매일 먹는 밥을 주로한 정식처럼 말이다. 느끼하지도, 자극적이지도 않으며, 몸에 편안한, 자연 본연의 맛을 느낄 수 있도록. 건축이 밥이라면, 삶은 찬과 같다. 한정식 찬의 수만큼 다양한 맛의 향연을 이루는 도시에서의 삶을 기대한다. 그리고, 지금은 밥 없이는 성립되지 않는 한정식 같은 건축이 필요하다. 사람이 거주할 거처가 필요하다.

2. 태도가 양식이 되다
"흔히들 '태도가 양식이 된다'라고 합니다. 전설적인 큐레이터 헤랄드 제만이 1969년 베를린 현대미술관에서 연 전시 'When Attitudes Became Form'은 현대미술 역사에 있어 작품뿐 아니라 전시의 언어를 전적으로 바꾼 역사적 순간이 되기도 했습니다. 건축, 도시, 동네를 논하면서, 태도라는 다소 동떨어진 단어가 어색할지도 모르겠습니다."

태도는 나의 오랜 친구, 홍보라 님과 동네이야기를 할 때 그녀가 선택한 단어였다. 그녀가 나를 표현한 문장을 인용한다.

"I've known Seungmo Seo as a neighborhood friend, a colleague, and an artist we worked together with for almost ten years now. Whenever he interacts with people, I always sense a person with an appropriate distance and warmth. What I want to emphasize here is the idea of an appropriate distance and warmth, which might be exactly what a city, neighborhood, or village needs. Attitude may sound abstract, but this attitude, when transformed into the layered landscape of a neighborhood—its structure, form, and the accumulation of moments among its members—tells the story like hands and fingerprints of a single person. It speaks of a city created by users, changing with each moment and temperature, ultimately shaping the world into different atmospheres."

"When Attitudes Became Form," mentioned above, historically represented the unique and beautiful role of art. It pursued absolute beauty, seeking to embody or reproduce that form. However, it now breaks free from those constraints, revealing how contemporary art takes shape based on each artist's unique attitude and social concerns, responding to the era and the environment of the city. Similarly, architecture, like fine art, has shifted its focus from the inherent form and materials of buildings and new attempts to a more comprehensive context within the city. Architects now face the dual challenge of reflecting or considering the physical and cultural context of buildings within the broader urban framework. Perhaps architecture has come to value 'relationality' more, reflecting a double burden placed on architects. However, this relationality, too, manifests in forms and structures based on each architect's unique interpretation and attitude."

— From an interview with Hong Borah

"제가 서승모씨를 동네친구이자 동료로, 또 함께 전시도 만들었던 작가로 알고 지낸 지 이제 거의 10년이 되어가네요. 사람들을 대할 때 항상 적당한 거리와 온도감을 가진 매우 개인적인 분이라는 느낌이 듭니다. 제가 여기서 주목하고자 하는 것은 적당한 거리와 온도감으로, 이것이야말로 도시, 동네, 마을에 필요한 것 아닐까? 생각합니다. 태도는 추상적으로 들릴지 모르지만, 이러한 태도가 동네의 구조, 형태, 구성원 간의 시간이 겹겹이 쌓인 풍경으로 치환되어, 마치 한사람의 손과 손금처럼, 시간이 적층된 풍경을 이야기합니다. 사용자에 의해 변경되는, 각각의 시간성과 온도감을 가지고 만들어지는 도시. 그렇게 해서, 세계의 유수는 각기 다른 분위기로 만들어집니다."

"위에서 언급한 'When Attitudes Became Form'은 역사적으로 미술의 역할이 고유하고 아름다운 형태, 즉 절대미를 추구하며 그 형태를 구현하거나 재현하는 것이었다면 이제 그 굴레에서 벗어나 예술가 각자가 가진 고유의 태도와 사회적 고민이 그 시대와 도시라는 환경에 어떻게 반응하는가에 따라 현대미술의 형식이 되는 현상과 여러 시도들을 보여주었습니다. 건축 역시 순수미술처럼 건축물의 형태와 재료의 고유성, 새로운 시도에 방점을 찍었던 것에서 좀 더 도시의 전체적 맥락 안에서 건물이 가지는 물리적, 문화적 맥락을 모두 반영하거나 고려하는 이중 부담(?)이 건축가들에게 부여된 것 같습니다. 건축이 좀 더 '관계성'을 중요시 여기게 되었다고 할까요. 그런데 그 관계성 역시 각자 건축가가 해석하는 고유의 방식과 태도에 의해 형태로, 또 형식으로 드러납니다."

— 홍보라와의 인터뷰 중에서

3. From Games…

In a context of rapid economic growth and population increase, urban planning methods inspired by games might have been effective. However, in the current situation of population decline and cities already filled to capacity, the conventional methods of urban planning and architectural design are still effective. The idea of drawing lines and creating utopias with an attitude similar to pure art might be an ignorant endeavor. Cities are now multi-layered complexes. In the ever-changing conditions of today, long-term plans like five-year plans appear ineffective. The meaning of concepts like historic-cultural zones, residential areas, and commercial zones may feel distant. While numerous plans are being developed, they also risk being overturned. Plans are discarded, altered, and shaken in the face of constant changes.

3. 게임으로부터

경제가 고도성장하고, 인구가 증가하는 근대상황에서의 도시계획방법론은 유효했을지도 모른다. 그러나 인구가 감소하고, 건물로 도시가 가득찬 지금, 구태의연한 도시계획과 건축설계의 방법론이 여전히 유효할까는 의문이다. 이런 상황에서 순수미술을 하듯, 유토피아를 만들겠다는 태도로, 선을 긋는 행위는 무지한 일일 것이다. 현재, 도시는 다층적인 복합체이다. 조건, 상황이 시시각각 변하는 지금, 무슨 5개년 계획 같은 것은 유효하지 않고, 역사문화지구, 주거, 상업지역 같은 의미 역시 멀게만 느껴진다. 무수한 계획이 수립되는 동시에 넘어지기 일쑤이다. 그렇게 계획은 폐기되고, 변화하고, 흔들린다.

Games come to life as users navigate through their creators' rules and freely enjoy the experience. Similarly, architects establish plans for cities and buildings, while a diverse group of people engages and enjoys life within urban and architectural spaces. In our studio, we aimed to discover new urban and architectural methodologies through the analysis and critique of various games. For instance, in the game of Go, if the grid represents the city and black and white stones symbolize architecture, occupied territories fluctuate in a dynamic manner, much like programs evolving as organic entities. Architects are accustomed to top-down decision-making, but cities often deviate from planned trajectories. Torre David is a project that highlights a failed construction project in the heart of the city, exploring how users can form a community in such a space. We select a specific area in the city center and illustrate, through drawings, how users can enjoy the city and architecture from their perspective. Rather than an abstract and utopian city, it resembles a dystopian image from a movie, where warmth is present, and a new city coexists with the existing one. It is a city with human warmth and joy. From each critic's visual language, we hope to find new possibilities for urban and architectural plans.
While there may be a preconception that cities lack nature, they indeed host a unique ecosystem of fauna and flora as land, water, and air coexist. Our critical language, attitude, and temporally expressed media must invariably include a global ecosystem.

4. Epilogue
The holistic forms shaped by the gaze upon the Earth, city, town, and neighborhood, and the framework that contains them in a new city, are possible and worth looking forward to, created with deep insight.

Looking towards the first steps of winter.

게임은 만든이와 룰을 넘나들며, 자유롭게 즐기는 사용자로 성립된다. 건축가가 도시, 건축의 계획을 수립하고, 불특정한 다수가 도시, 건축의 공간에서 삶을 영위하고 즐기는 것과 같다. 그래서 우리 스튜디오에서는 다양한 게임의 분석, 비평과정을 통해 새로운 도시, 건축방법론을 찾으려 했다. 예를 들면, 바둑에서 그리드가 도시라면 흑과 백은 건축이고, 점유된 영역이 흑과 백의 상황에 따라 변화하듯, 현재적으로 프로그램은 성장하는 유기체이다. 건축가는 탑다운 의사결정에 익숙하다. 그러나, 현재 도시를 보면, 계획대로 흘러가지 않는 경우가 다반사이다. 토레다비드는 도시 심장부에 버려진 실패한 건설 프로젝트지만, 사용자들이 어떻게 공동체를 형성할 수 있는가?를 보여주는 단적인 프로젝트이다. 우리는 구도심의 한 지역을 선택하고, 사용자의 관점에서 도시와 건축을 즐기는 방식을 드로잉으로 표현한다. 추상적이고 절대적인 이상향의 도시라기보다는 온기를 머금은 이들의 도시에 가까운, 신구가 공존하는 영화 속 디스토피아적인 이미지에 가까울 것 같다. 그럼에도 인간미가 살아있고, 즐거운 도시 말이다. 각자의 비평적 언어가 시각화된 그것으로부터 새로운 도시, 건축 계획에 한발 더 다가갈 수 있는 가능성을 찾을 수 있기를 희망한다. 도시에는 자연이 없는 듯한 선입견을 갖지만, 땅과 물, 대기가 공존하는 한 그 나름의 지구생태계의 동식물이 같이 한다. 우리의 비평적 언어, 태도, 시간화된 표현 매체에는 반드시 지구 규모의 동식물계가 포함되어야 한다.

4. 에필로그
지구, 도시, 마을, 동네를 바라보는 시선, 깊은 눈빛으로 만들어진 형상들의 총체, 그리고 이를 담는 새로운 도시의 틀은 가능하며, 충분히 기대할 만한 것이라 생각한다.

겨울의 첫걸음을 바라보며.

Before You Read — Calvin Chua

Can urbanism be drawn from the profiles and characters of our city neighborhoods?

The '15-minute city' has been a trending urbanism concept for the past two years. Since the announcement of its adoption by the city of Paris in early 2020, the '15-minute city' has become a reality for most people during the pandemic where the living environment has been reduced to a minimal commuting radius. The '15-minute city' envisions a model of decentralised multi-purpose neighbourhoods containing all the required amenities as a strategy to reduce carbon emissions, while improving liveability. Such a concept is not new; from Cerda's Barcelona grids to Moscow's Mikrorayons, the '15-minute city' has existed in different urban forms.

Even within Seoul, new mixed-use residential neighbourhoods, such as Sewoon Grounds and H1, have promised the integration of social and work amenities within the development, shrinking the need for long distance commutes, while seeding the possibility of jump-starting localised social and economic development. While these projects are ambitious in rethinking the future of neighbourhoods, they are only currently possible through the redevelopment of idle brownfield sites or replacing existing neighbourhoods to achieve the required density.

Therefore, the central interest of our studio is to rethink the approach towards the 15-minute city through the lens of strategic intensification of existing neighbourhoods in Seoul through a series of 'urban rooms'.

우리의 도시—동네의 프로필과 캐릭터로부터 어바니즘을 끌어낼 수 있는가?

'15분 도시'는 지난 두 해 동안 주목받는 도시 계획 개념이다. 2020년 초 파리가 채택을 발표한 이후 '15분 도시'는 팬데믹으로 인해 거주 환경이 최소 통근 반경으로 축소된 상황에서 대부분의 사람들에게 현실이 되었다. 탄소 배출을 줄이고 삶의 질을 향상시키기 위한 전략으로 모든 필수 시설을 포함하는 분산된 다목적 동네 모델의 개념은 새로운 것이 아니다. 세르다의 바르셀로나 그리드부터 모스크바의 미크로레이온까지 '15분 도시'는 다양한 도시 형태로 존재해 왔다.

서울 내에서도 새로운 혼합 사용 주거 동네인 세운구도와 H1과 같은 프로젝트들은 사회 및 업무 시설을 개발 내에 통합하여 멀리 있는 통근 필요성을 줄이고 지역사회 및 경제 발전을 촉진하는 가능성을 제시하고 있다. 이러한 프로젝트들은 동네의 미래를 재고하는 데 야심차지만, 필요한 밀도를 달성하기 위해 텅 빈 토지를 개발하거나 기존 동네를 대체하는 것만이 가능하다. 그래서 우리의 중심적인 관심사는 '도시의 방'이라는 일련의 전략적 집중을 통해 서울의 기존 동네에 대한 15분 도시 접근 방식을 재고하는 것이다.

Conceived as the basic building block of the city, the 'urban room' can be found at two extreme scales. At the micro scale, the naming of rooms or 'bang' is culturally tied to social spaces and amenities within the city - think of dabang, jimjilbang, noraebang, etc. At the macro scale, the definition of urban rooms is tied to the mega structuralist tendencies of the 1960s, characterised by large interiorised social and commercial spaces.

Working between two scales, the goal is to develop projects for selected neighbourhoods, benefiting 2000 inhabitants. By shifting the definition of inhabitants away from dwelling units to beneficiaries, it opens up possibilities for spatial interventions that respond to the specificities for the neighbourhood. Also, instead of applying a generic layer of mixed-use, residential and commercial programs, we capitalise the very essence of the neighbourhood's character and profile to define our spatial interventions and define a future vision for the neighbourhood's transformation.

도시의 기본 구성 요소로 고안된 '도시의 방'은 두 가지 극단적인 규모에서 찾을 수 있다. 미시적 규모에서는 방이나 '방'의 명칭이 도시 내의 사회 공간 및 편의 시설과 문화적으로 연결되어 있다. 대표적으로 다방, 찜질방, 노래방 등이 있다. 거시적 규모에서는 도시의 방을 정의하는 것이 1960년대의 거대한 내부화된 사회 및 상업 공간에 대한 메가 구조주의적 경향과 관련이 있다. 두 규모 사이에서 작업하며, 2000명의 주민을 지원하는 특정 동네를 위한 프로젝트를 생각해 본다. 주민의 정의를 주거 단위에서 수혜자로 바꾸면 동네의 특이성에 대응하는 공간 개입의 여지가 열릴 수 있다. 또한 일반적인 혼합 사용, 주거 및 상업 프로그램의 일반적인 층위를 적용하는 대신, 동네의 본질과 사용자 프로필을 활용하여 우리의 공간 개입을 정의하고 동네의 변화에 대한 미래 비전을 제시한다.

파트 2.
Part 2.

Why 'Keywords, Places, Purposes'?

In terms of the perspectives embodied in our architecture:
— What does the project represent?
— What do we individually value?
— What shared vision do we express through the keywords?

Concerning the places associated with our architecture:
— How significant is the notion of place within the project?
— Can it exist anywhere, or does it exist nowhere?
— As we explore these places, what insights do we derive?

Regarding the purpose inherent in our architecture:
— Does the absence of purpose or the introduction of a new purpose create spatial freedom?
— What needs to be changed?
— What purposes are we collectively willing to embrace?

왜 '키워드, 장소, 용도'인가.

우리의 건축이 담는 관점에 대하여
— 무엇을 프로젝트가 가리키는가.
— 무엇을 중요하게 생각하는가.
— 무엇을 우리가 함께 바라보는가.

우리의 건축이 연관된 장소에 대하여
— 장소가 프로젝트에서 얼마나 중요한가.
— 어디에도 있을 수 있는가,
  어디에도 있지 않은가.
— 장소를 가로지르며, 우리는 무엇을 도출하는가.

우리의 건축이 대응하는 '용도'에 대하여
— 용도없음, 또는 새로운 용도는 공간적 자유를 주는가.
— 무엇을 바꾸어야 하는가.
— 우리는 무슨 용도를 함께 할 것인가.

	강하림 Harim Kang	김보경 Bogyong Kim	김서진 Seojin Kim	이연우 Yonu Lee
땅 Land	●		땅으로서 도로 Road as a Land ●	●
			●	
인프라 Infrastructure	●	시간성에 기반한 경계 Temporal Boundaries ●	●	●
			환경 인프라 ● Environmental Infrastructure	●
공공 Public	●	파사드 Façasde ●	●	●
		구축 ● Construction		●

읽는 방법 1   김태영

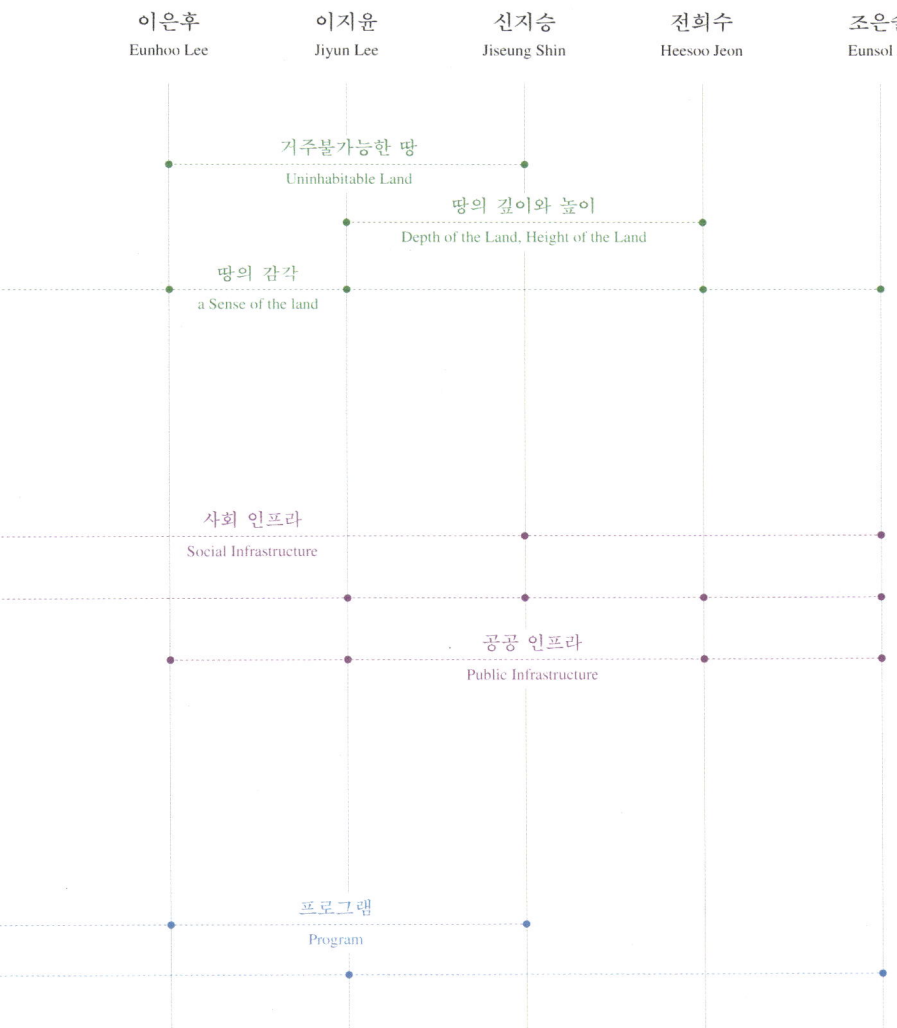

Keywords:
Land, Infrastructure, Public

Reflecting on the graduation projects of the nine individuals, the major keywords that emerge are "Land, Infrastructure, Public." While these keywords may appear common, their interpretations and attitudes vary depending on the studio and the individual. Some projects delve deeply into a single keyword, while others adopt a broader perspective by incorporating all three.

Examining the projects through the lens of these keywords reveals a synergy where seemingly disconnected works can be understood in relation to one another. This approach broadens our awareness of issues that extend beyond individual projects, occasionally leading to a more in-depth exploration of a single keyword and, at other times, offering a broader perspective by traversing all three. Focusing on these keywords prompts us to consider our collective vision and what is deemed essential. The potential for shared thinking and collaborative action serves as a means of reconnection.

키워드:
땅, 인프라스트럭쳐, 공공

2022년 한국예술종합학교 건축과 졸업 작품인 9인의 작업에 대한 회고와 성찰을 통해, 주요 키워드로서 '땅, 인프라, 공공'을 읽는다. 이는 공통된 키워드일뿐, 해석이나 태도는 스튜디오에 따라, 개인에 따라 다르고 다양하다. 하나의 키워드를 더욱 깊이있게 다루는 작업도 있고, 이 세 키워드를 가로지르며 더 넓게 조망하는 작업도 있다.

키워드를 중심으로 바라본다면, 연결되지 않았던 작업들이 함께 읽히기도 하고, 개인의 작업을 넘어서 공동의 아젠다로 문제의식을 확장하기도 한다. 개별 작업의 고유함을 유지하면서도 우리가 무엇을 함께 바라보고 있는가, 무엇이 중요한가를 생각하게 한다.

함께 사고할 수 있는 가능성, 함께 실천할 수 있는 가능성이 우리를 다시 연결한다.

1 Uninhabitable Land as New Land

The inquiry into land begins with an awareness and concern that the conditions of buildable land do not necessarily lead to good architecture. As we observe the increasing scarcity of desirable land and consider the nullification of habitability in seemingly permanent land due to climate change, disasters, and catastrophes, it is natural to question whether our urban architecture should be based solely on existing land. Just as modern urban housing, supplied on a large scale, appears indifferent to fundamental issues such as "the unhomely" or "uninhabitable" conditions in contemporary society and cities, the indifference of the conditions provided by modern cities prompts us to seek the freedom to create new urban architecture from land that currently seems uninhabitable.

Eunhoo Lee's work challenges the assumption that habitable land is a prerequisite for habitable architecture. The outcome is a public building situated on a waterlogged site, surrounded by and adjacent to it. Another work exploring uninhabitable land is by Jiseung Shin. In a society that produces "useless" individuals, acts of art and contemplation illustrate the process of finding value in these so-called useless humans by inhabiting uninhabitable spaces. Transforming uninhabitable land into habitable land represents a society where individuals seeking new ways of living come together. The architecture that encapsulates this notion resembles a mountain—a mound—where life inside is akin to a window facing the mountain, filled with light penetrating deep into the ground, piercing through the mountain's surface, casting clear shadows.

1 새로운 땅으로서 거주불가능한 땅

땅에 대한 질문은, 건축 가능한 땅이 가진 조건들이 반드시 좋은 건축으로 이끄는 전제가 되지 않음에 대한 인식과 문제의식으로부터 시작한다. 모두가 살고 싶어하는 땅이 더 더욱 희소해지는 것을 목격하고, 기후변화나 재난, 참사가 영원할 것 같던 땅의 거주가능성을 무효화시키는 것을 바라보며, 우리의 도시건축이 기존의 땅을 전제로 하여야 하는가에 대한 질문을 가지는 것은 어쩌면 당연하다 . 대규모로 공급되는 도시 주거가 현대사회, 현대 도시에서의 '집없음,' '거주불가능함'과 같은 근본적인 문제를 해결하는데 무관심하듯, 현대 도시가 공급하는 땅의 조건 또한 건축을 세우는 근본적인 토대, 우리가 공동으로 가져야 할 땅을 통한 연대에 무기력하다면, 우리는 눈을 돌려 지금은 거주불가능해 보이는 땅으로부터 새로운 도시 건축을 가능케 할 자유를 찾는다.

우리에게 땅은 영원한가라는 근본적인 질문으로부터, 집을 지을 수 있는 땅이 거주가능함을 만드는 필연조건이 아니라는 것을 이야기하는 이은후의 작업이 있다. 집을 지을 수 없는 땅, 거주할 수 없는 땅이 오히려 새로운 집, 새로운 건축을 제시할 수 있는 조건을 만든다고 전제한 그 결과는 유수지 위의, 유수지를 둘러싼, 유수지 옆의 공공건축이다. 거주할 수 없는 땅으로부터 시작하는 또 다른 작업으로 신지승의 작업이 있다. '무용'의 인간을 만드는 사회에서 성찰과 예술과 같은 무용의 행위가, 거주불가능한 곳에서도 거주가능한 집을, 그 안에서의 거주로 무용의 인간이 가지는 가치를 찾는 과정을 보여준다. 거주할 수 없는 땅을 거주할 수 있는 땅으로 만드는 것은, 함께 사는 방법을 새롭게 찾아가는 이들이 모인 사회이다. 그것을 담는 건축은 산을 닮았고, 구릉을 닮았고, 그 안의 삶은 그 산을 바라보는 창, 산의 표피를 뚫고 땅속 깊이 들어오는 빛이 가득하고 그림자가 뚜렷하다.

Jiyoon Lee employs the traditional method of blending air with the land to propose public architecture as a constructive approach aimed at healing the historical wounds embedded in the land. This public architecture serves as a proposition for the restoration of shared land, signifying the beginning of reclaiming its identity through the recovery process. Heesoo Jeon's work suggests an infrastructure to cool the land made inhospitable by the heat island effect, incorporating a roof that serves as a public park. In between, temporary residences are planned, presenting a public skyscraper accessible to all.

Both individuals demonstrate how, despite being land that is universally desired or of historical significance, ownership or occupation can hinder the communication of that value, leading to specific architectural outcomes or confining it to an exclusive environment. Each proposal invites a renewed perspective on the depth and elevation of the land, underscoring the necessity for inclusivity.

Conversely, the works of Harim Kang, Bogyong Kim, and Yonu Lee challenge the notion that reserving roads as land where no construction is permitted is the optimal approach for ensuring public accessibility. They suggest that when architecture operates under the premise of temporality, utilizing roads as land can contribute to better urban boundaries.

반면, 우리가 익숙하게 거주할 수 있다고 생각하는 땅에 대한 질문이 있다. 땅에 공기를 섞는 전통적 방식을 이용해 땅이 가진 역사적 상처를 치유하는 구축의 방법으로서 공공 건축을 제안하고, 그렇게 회복된 함께 디디고 있는 땅이 정체성을 회복하는 시작임을 지시하는 이지윤의 작업과, 열섬 효과로 살기 어려워진 땅을 식히는 쿨링 타워와 그 지붕을 공공을 위한 새로운 땅으로 제안하고, 그 사이를 임시 주거로 계획하여 공공의 마천루를 제안하는 전희수의 작업이다.

두 사람은 모두가 살고 싶어하는 땅이거나 역사적으로 의미있는 땅임에도 그것의 소유나 점유가 그 가치의 소통을 배제하였을 때 어떤 건축물로 귀결될 수 있고 어떤 배타적인 환경으로 남겨질 수 있는지를 드러낸다. 각각의 제안이 땅의 깊이 또는 땅의 높이를 새롭게 바라보고 포용할 수 밖에 없는 이유이다.

한편으로, 공공성을 확보하기 위해 누구도 건축할 수 없는 땅으로 전제하는 도로에 대해 그 유보가 최선인가를 질문하고, 시간성을 전제로 작동하는 건축이 도로를 대지로서 점용할 때 더 좋은 도시 경계부를 만들 수 있음을 제안하는 강하림, 김보경, 이연우의 작업이 있다.

2 Infrastructure Expanding Architecture

Our city is remarkably generous towards infrastructure. Whether it involves high-speed trains, highways, overpasses, interchanges, or seawalls, the execution of these projects is swift once agreements and procedures related to the values they create—such as public access, the convenience of material movement, regional and urban connectivity, and the asset value of adjacent land—are completed. The impact on the landscapes of villages maintained for centuries, changes to the natural environment, and effects on residential settings are often accepted through compensatory measures such as the creation of new public spaces or the installation of soundproof barriers. Perhaps creating a better city requires expanding the value and role of radical architecture by incorporating elements of urban, environmental, or social infrastructure.

Seojin Kim's work reorganizes architecture and public spaces as surfaces allowing rainwater permeation, while Eunsol Jo's project transforms the vibrations and sounds of people walking on subway platforms into a shared sensory experience. Both suggest the potential of architecture as a social and environmental infrastructure. Eunhoo Lee's work combines public architecture with reserved land designed to safeguard urban areas from flooding, creating land that remains viable even in times of crisis. Jiyoon Lee's project restores the ecological system of the land through the integration of air, and Heesoo Jeon's work addresses the heat island effect through cooling infrastructure. Together, these projects illustrate how architecture can meet the challenges posed by the land and serve as infrastructure that fosters a healthier environment.

2 건축을 확장하는 인프라스트럭쳐

우리의 도시는 인프라스트럭쳐에게는 매우 관대하다. 고속철도, 고속도로, 고가도로, 인터체인지, 방파제… 거대하거나 매우 빠르거나 극단적으로 높아도 그것이 만드는 가치 — 공공성, 물자 이동의 편리함, 지역의, 도시의 연결, 인접대지의 자산 가치 등 — 에 대한 합의와 절차만 완성되면 집행은 신속하다. 몇 백년 유지되어 온 마을의 경관, 자연 환경의 변화, 거주 환경에 미치는 영향은 새로운 공공 공간을 통한 보상, 방음벽 설치 등을 통해 받아들여질 수 있다. 어쩌면 더 나은 도시를 만들기 위한 급진적 건축은, 도시적, 환경적, 또는 사회적 인프라스트럭쳐의 영역을 포함함으로써 그 가치와 역할을 확장함으로써 가능할 수 있다.

우수를 흘려보내는 지표면으로서 건축물과 공공 공간을 재조직하고 식물을 돌보는 경계부 인프라로 주거를 재조직하는 김서진의 작업, 지하철 또는 그 안을 걷는 이들의 진동과 소리를 함께 공유할만한 감각적 경험으로 바꾸는 조은솔의 작업은 사회적, 환경적 인프라스트럭쳐로서 건축의 가능성을 제시한다. 범람과 침수로부터 도시의 땅을 지키기 위해 유보된 땅인 유수지가 그 어떤 위기에도 사라지지 않는 땅으로서 공공건축과 결합되는 이은후의 작업, 땅의 생태계를 공기로 회복시키는 이지윤의 작업, 우수를 쿨팅 인프라로 열섬을 식히는 전희수의 작업 또한 땅에 대한 도전으로부터 환경을 만드는 인프라스트럭쳐로서의 건축으로 이어지는 작업이다.

One project redefines both architectural façades and interiors as boundary infrastructure that operates based on user mediation and appropriation, emphasizing their temporality.

Harim Kang's work reinterprets architecture as urban infrastructure by adapting usage patterns and construction methods, thus presenting new urban spaces and lifestyles. Questioning the traditional social institution of a wedding and its associated architectural stereotypes, Bogyong Kim's work introduces an open and inclusive structure that expands into the city, fostering a temporary sense of unity among anonymous city dwellers. Additionally, Yonu Lee's project suggests the façade of the market building as a boundary infrastructure that encourages mediation and consensus among different users and residents surrounding Dongmyo Market.

Their commonality lies in the attempt to combine social infrastructure—institutions, facilities, and communities that have sustained societal functions—with urban infrastructure adapted to the concept of temporality.

The works of Harim Kang and Yonu Lee, which foster communication, coordination, and mediation among neighbors; Seojin Kim, who promotes a care community mediated by plants; Jiseung Shin, who encourages collective reflection and creation; and Eunsol Jo, who shares individual journeys through sound and energy—all converge towards a common goal: to propose a "social infrastructure" that reflects the relationships between individuals, neighbors, and communities.

사용자 간의 조정과 중재, 그들의 시간성에 기반해 작동하는 도시-건축 경계의 인프라스트럭쳐로서 건축 파사드와 내부공간을 재정의하는 작업이 있다.

임시적인 교환과 교류를 독려하는 경계부가 공과 사의 영역 사이에서 만드는 새로운 도시공간, 새로운 도시일상을 제시하는 강하림의 작업은 달라질 수 밖에 없는 점용의 방식, 구축의 방식으로 도시적 인프라스트럭쳐로서 건축을 재정의한다. 결혼식이라는 특수한 사회적 제도와 이와 연관된 제한된 건축유형에 대해 질문하고, 이를 개방과 포용의 구축성을 가진 구조물로 덧붙여 도시로 확장함으로써 익명의 도시인들 간에 일시적 유대감을 만드는 김보경의 작업이 있다. 동묘시장을 둘러싼 서로 다른 사용자, 거주민 간에 조정과 중재, 합의를 유도하는 경계장치로서 건축물의 파사드, 공과 사의 경계부를 제안하는 이연우의 작업이 있다.

이들의 공통점은, 사회의 기능을 유지해 왔던 제도, 시설, 공동체를 만드는 사회적 인프라스트럭쳐를, 시간성에 따라 조정되는 도시적 인프라스트럭쳐와 결합하려 한 시도이다.

강하림과 이연우(이웃의 소통/조정/중재), 김서진(식물을 매개로 한 돌봄공동체), 신지승(성찰과 창작을 공유하는 개인의 집합), 조은솔(이동하는 감각을 소리와 에너지로 공유하는 개인들의 여행)의 작업은 개인과 이웃, 공동체의 관계에 대한 각자의 제안을 '사회적 인프라스트럭쳐'를 제안하고자 한 궁극적인 목표가 있다.

3 Suggestions for Public Realms

As a new foundation for urban architecture, the proposal to explore the potential of uninhabitable land and the infrastructure extending from architecture demands a fresh approach to communal living. At its core, this requires a reassessment and new proposals for public spaces, addressing how we inhabit these spaces, what functions they should serve, and what architecture can encapsulate them.

Eunsol Jo's work presents transportation infrastructure as a new form of public architecture, aiming to highlight the importance of movement for humanity by envisioning alternatives to conventional efficient and convenient means of connecting origins and destinations. Heesoo Jeon's work illustrates the necessity for public architecture, such as skyscrapers, to serve as a powerful infrastructure that matches the scale of environmental inequalities exacerbated by economic disparities. Jiyun Lee's work goes beyond individual buildings and substructures to propose a new ecosystem, thus broadening the architectural spectrum.

3 공공 영역에 대한 제안

도시건축의 새로운 토대로서 거주불가능한 땅의 가능성과 건축을 확장하는 인프라스트럭쳐의 제안은, 함께 사는 방식에 대한 새로운 접근방식을 요구한다. 그 중심에는 공공의 영역을 어떻게 전제하고 어떤 기능을 담당하며 어떤 건축으로 그를 담을 수 있는가와 같은, 공공 영역에 대한 재고와 새로운 제안이 있을 수밖에 없다.

교통 인프라를 새로운 공공 건축의 유형으로 제시하는 조은솔의 작업은, 출발지와 목적지를 잇는 효율과 편리함의 수단이 아니라 삶의 대안을 꿈꾸고 탐색하는 인류에게 이동의 중요성을 깨닫게 하는 목적 자체가 될 수 있음을 보여주고자 한다. 전희수의 작업은, 경제적 불평등이 심화시키는 환경적 불평등은 그 문제의 스케일만큼 강력한 인프라로 작동할 수 있는 공공 건축 - 마천루 - 가 필요함을 보여준다. 단일 건물 또는 하부 구조를 넘어서 새로운 생태계를 제안하려는 이지윤의 작업은 그 스펙트럼을 넓힌다.

Eunhoo Lee's work combines public programs that are not solely predicated on the land, while Yonu Lee's project draws exchanges and interactions from private areas into public spaces, emphasizing the complexity and expandability of public initiatives. The efforts of Bogyong Kim, Eunhoo Lee, Jiyun Lee, and Eunsol Jo seek to imbue public spaces—free from specific characters, atmospheres, or programmes—with new intent, aiming to find frameworks and surfaces that provide purpose and functionality. They strive to reveal the characteristics and techniques of materials through geometry and propose constructiveness.

The works of Harim Kang, Yonu Lee, and Seojin Kim envision a community that extends the boundaries between public and private realms, fluidly navigating ownership and occupancy, exchange and interaction, collaboration, and care. These projects challenge the limitations of current methods that separate and define public and private domains, thereby restricting the permissions and regulations governing architectural actions. In urban living, the consideration of publicness becomes imperative, regardless of property ownership. Seojin Kim, Jiseung Shin, and Heesoo Jeon address the challenges inherent in housing, acknowledging the need to confront issues related to exploring better ways of living together and alternative modes of existence. Their works integrate housing as a public good with social and environmental infrastructure.

그 땅에 전제되지 않았던 공공 프로그램을 복합하는 이은후의 작업이나, 사적 영역에서 이루어지는 교환과 교류의 프로그램을 공공 공간으로 끌어내려는 이연우의 작업은 공공 프로그램의 복합이나 확장 가능성에 주목한다. 캐릭터나 분위기, 프로그램이 특정되어서는 안 된다고 믿는 공공 공간에 대해, 새로운 의도를 부여하고 그것을 작동하게 하는 뼈대와 틀, 면을 찾으려는 김보경, 이은후, 이지윤, 조은솔의 노력은, 재료의 특성과 구법을 드러내는 지오메트리, 구축성을 제안하고자 한다.

공공의 영역을 공과 사의 경계 영역으로 확장하고, 소유와 점유, 교환과 교류, 협력과 돌봄의 경계를 느슨하게 오가는 공동체를 제안하는 강하림, 이연우, 김서진의 작업은, 오늘날 공적 영역과 사적 영역을 구분하고 정의하는 방식, 건축 행위의 허용과 규제를 제한하는 방식의 한계에 도전한다. 마지막으로, 도시 주거는 이웃을 만나고 환경을 경험하며 도시 일상을 만들어 간다는 점에서 주택의 소유와 무관하게 공공성을 고려할 수밖에 없다. 김서진, 신지승, 전희수의 작업은, 함께 사는 더 나은 방식, 다른 방식을 탐색하는 과정이 수반할 수밖에 없는 주거의 문제를 직시하며, 공공재로서의 주거와 사회적, 환경적 인프라스트럭쳐를 결합한다.

How to read 1: Place and Purpose

Infrastructure	Common Space	Institution	Residence
도시기반시설 교통 인프라	공공공간 도로 공 – 사의 중간지대	제도	주거

읽는 방법 1: 장소와 용도

서울시 강북구	●● 김서진	Seojin Kim　　/ *Rain Festival*
		번동 주공아파트 3단지
		오현로 25가 – 라길 일대
		/ 저층고밀주거단지, 번동 임대아파트 단지

서울시 동대문구　　●●● 강하림　　Harim Kang　　/ *Cul-De-Sac*
　　　　　　　　　　　　　　　　천장산로4길, 이문로9길 일대
　　　　　　　　　　　　　　　　/ 현 다세대주택 밀집구역, 대학 인근 동네

서울시 종로구　　●● 이연우　　Yonu Lee　　/ *Dongmyo: Intergenerational Network*
　　　　　　　　　　　　　　　청계천로, 종로56길, 종로58길, 지봉로2길,
　　　　　　　　　　　　　　　지봉로4길, 난계로25길
　　　　　　　　　　　　　　　/ 현 동묘시장 일대

서울시 중구　　●● 김보경　　Bogyong Kim　/ *Magnetic Field City*
　　　　　　　　　　　　　　퇴계로, 필동로 일대
　　　　　　　　　　　　　　/ 결혼식장과 인접 도로, 공공 공간

서울시 용산구　　● 이은후　　Eunhoo Lee　　/ *The Fourth Attack: Linking Instabilities*
　　　　　　　　　　　　　　한강대로 11길 / 한강대교 북단 빗물펌프장

　　　　　　　　● 이지윤　　Jiyun Lee　　/ *Plowing and Blowing*
　　　　　　　　　　　　　　한강대로 160 / 현 한국전력 창고부지

　　　　　　　　● 조은솔　　Eunsol Jo　　/ *Dance Rock-et: generating metro*
　　　　　　　　　　　　　　한강대로 지하185 / 현 삼각지역

서울시 강남구　　●● 전희수　　Heesoo Jeon　　/ *Tip-Toeing Landscape*
　　　　　　　　　　　　　　논현1동 일대
　　　　　　　　　　　　　　/ 강남 블럭 중 하나의 중심, 저층고밀주거단지

서울시 강서구　　●● 신지승　　Jiseung Shin　　/ *The Mounds, Artifacts*
경기도 고양시　　　　　　　　강서구 양천로 201-1,
　　　　　　　　　　　　　　경기도 고양시 덕양구 대덕로 426
　　　　　　　　　　　　　　/ 현 한강의 남측 및 북측 천변에 위치한
　　　　　　　　　　　　　　　물재생센터

메인 드로잉
Main Drawing

제목 Title

키워드
Keyword

사이트
Place

유형
Purpose

내용
Content

The index card provides a quick preview of the students' work, organized by keyword, place, and purpose, as well as the main drawing.

The main drawing, title, and content are the same as those posted on the previous graduation exhibition website, and the 10 keywords that describe each student's work give a glimpse into the perspective and value of each student's work.

By categorizing the students by studio, it is possible to see the individual works analyzed in the previous part, Part 2.

인덱스 카드는 학생들의 작업을 키워드, 사이트, 유형 세 개의 요소와 메인 드로잉으로 미리 간략하게 읽을 수 있는 파트이다.

메인 드로잉과 제목, 내용은 기존의 졸업 전시 홈페이지에 게재된 사항들과 동일하며 학생별로 본인의 작업을 설명하는 10개의 키워드를 통해서 각 학생들이 주목한 작업의 관점과 가치를 들여다 볼 수 있다.

스튜디오별로 학생들을 분류해 앞선 Part 2. 읽기 전에를 통해서 항목별로 분석된 개인의 작업을 한 눈에 보는 동시에 그룹화를 통한 스튜디오의 특징을 다각도로 바라볼 수 있게 구성하였다.

Residents of Bun-dong, where permanent rental apartments were first introduced, cannot escape Bun-dong for economic and physical reasons. The urban organization and its natural topography, which began with the valley of the two mountains, made rainwater faster, and they were vulnerable to responding to such rainwater. I want to demand architectural changes to rain water so that I don't fear rain, and let their daily lives permeate the plants so that they wait for tomorrow when flowers bloom.

Architectural elements are transformed to cope with rainwater and fill the space between us with plants and water. Bring the things that plants need into the house. In a plant cycle, our senses rise again, and we can become one like a cluster of plants. Rainwater becomes the medium that brings the city together, and we now wait for rain.

Connected by water and plants, they become a new local community that cares for each other. They are weak when they are individuals, but they become powerful when they become a system through solidarity between buildings. Plant communities born of equal rainwater recovery are sensuous and alive.

Many will be the irrelevant beings, meaningless exis-tences. Only when they move away from production and labor, they will think about what the fundamental acts are.

In order to remain as a human, they want to live in a spectrum from the active life to contemplative life (from working of materializing to do nothing like a meditation). Assuming that survival is a secured future, the residential space will belong to this active-contemplative space.

For this life, they go to the land that has been discontinuously emerged by the change of time. The two huge but different lands at the top of the modernized and underground water treatment plants are folded ground facing each other with a river in between. Several mounds on each site become 'defined folds' from the waterfront to the city and depict a circular life between the two places.

Those who doubt their foundation will live on, and at the same time, in an artificial mound that makes the experiences of the land in daily life. The soil will be dug up and covered on the top of the mound, and the renovated interior space and mountainous surfaces on the top of the underground facil-ities will exist at the same time, and even new buildings will be the mounds. Because the persons are by the earth.

김서진
Seojin Kim

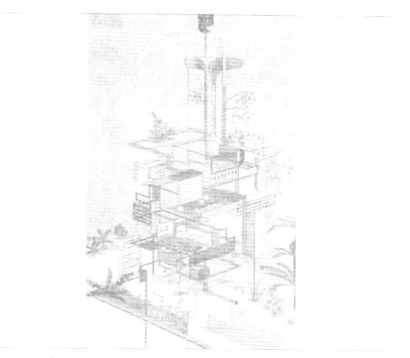

Rain Festival

키워드	도시재생, 리노베이션, 친환경, 사회불평등, 식물, 그린인프라, 영구임대, 프로토타입, 공동체, 돌봄
사이트	서울시 강북구 번동주공 아파트 3단지 서울시 강북구 오현로25 일대
유형	주거 리노베이션

영구 임대 아파트가 최초로 도입된 번동의 주민들은 경제적, 물리적 이유로 번동을 벗어날 수 없다. 두 산의 골에서부터 시작되는 도시 조직과 타고난 그 지형은 빗물을 더욱 빠르게 만들고, 그러한 빗물에 대응하기에 그들은 취약했다. 비를 두려워하지 않도록 빗물에 대한 건축적 변화를 요구하고, 그들의 일상을 식물들 속에 스미게 하여 꽃이 피는 내일을 기다리게 하고 싶다.

건축적 요소들은 빗물에 대응하기 위해 변형되고, 우리 사이를 식물과 물로 채워나간다. 식물에게 필요한 것들을 집 안으로 들인다. 식물의 사이클 속에서 우리의 감각은 다시 일어나고, 우리는 식물의 군집처럼 하나가 될 수 있다. 빗물은 도시를 하나로 만들어주는 매개체가 되고, 우리는 이제 비를 기다린다.

물과 식물로 연결된 그들은 서로를 돌보는 새로운 지역 공동체가 된다. 개인일 때는 미약하나 건물 간의 연대로 하나의 시스템이 될 때 그들은 강력해진다. 평등한 빗물의 회복으로 태어난 식물 공동체는 감각적이고 살아 있다.

신지승
Jiseung Shin

The Mounds, Artifacts

키워드	지형, 또 다른 지구, 공감각, 강, 같지만 다른 조건, 카펫, 마운드, 설계 순서, 통합적인 삶, 활동과 관조, 무용계급
사이트	서울시 강서구 양천로 201-1 경기도 고양시 덕양구 대덕로 426
유형	복합시설

주름의 인공산에 무용계급이 거주한다. 다수는, 의미 없는 존재인 무용계급이 될 것이다. 생산과 노동에서 멀어졌을 때 비로소 근본적인 행위가 무엇인지 생각해본다. 이들은 인간으로 남기 위해 활동적 삶부터 관조적 삶까지의 스펙트럼(물질로 실체화하는 제작부터 아무것도 하지 않는 명상까지)에 살고자 하게 된다. 생존을 확보된 미래로 가정하고, 이 활동-관조의 공간에 주거 공간이 종속된다. 이 삶을 위해, 기술 발전에 의해 불연속적으로 떠오른 땅으로 간다. 현대화, 지하화된 물재생센터 상부에 생긴 거대한 같지만 다른 두 대지는, 강을 사이에 두고 서로 마주보는 주름진 곳이다. 각 대지에 여러 마운드들이 강변부터 도시까지를 걸친 인공 주름이 되고, 두 곳을 오가는 순환적 삶을 그린다. 본인의 기반을 의심하는 이들은, 땅의 경험이 일상이 되게 하는 인공 마운드 위에서, 동시에 그 속에서, 거주할 것이다. 흙은 파내지고 마운드 상부에 덮어질 것이고, 지하 시설 위에는 개축된 내부공간과, 상부의 산 같은 표면이 동시에 존재할 것이며, 신축된 건물들조차 마운드가 될 것이다. 사람은 땅에 의한 존재기 때문이다.

조은솔
Eunsol Jo

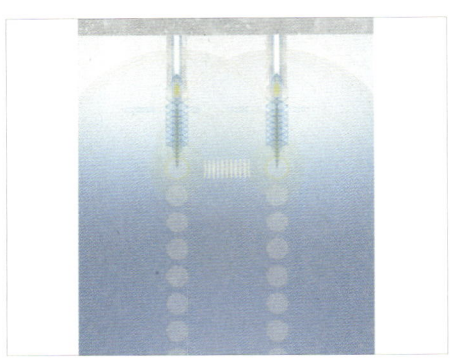

Dance Rock-et: Generating Metro

키워드	보편, 상상, 에너지, 이동, 인간, 여행, 진동, 춤, 지하철, 클럽
사이트	서울시 용산구 한강대로 1가 45-1 현 지하 삼각지역 일대
유형	복합시설

This project proposes a new type of platform that is recognized as a production activity by creating energy using the vibration and wind of the subway in Samgakji Station, which has a long isle for transfer purposes.

Humans are by nature moving. They had so many questions.

Movement itself was production for humans. By dreaming and anticipating the next.

Some of them lost their curiosity. And they named it as a responsibility, Settlement designed inequality.

'Imagination' should be equal to everyone as a universal 'productive activity' rather than a 'surplus' of privileged people.

본 프로젝트는 환승의 용도로만 이루어져 긴 유휴 환승통로를 가진 삼각지역에 지하철의 진동과 바람을 이용하여 에너지를 만들고 이곳에서의 움직임이 만들어내는 경험과 상상이 생산활동으로서 인정되는 새로운 형식의 플랫폼을 제안한다.

인간은 원래 움직이는 동물이었다. 호기심을 가진 인간은 상상하며 계속 걸었다. 움직임 자체가 생산이었다. 꿈을 꾸고 다음을 기대하면서. 호기심을 잃은 몇몇은 정착했고 그것을 책임이라고 이름 지었으며 서로를 나누는 차별로 이어졌다.

이 공간에 존재하는 우리는, 끊임없이 꿈꾸는 인간이어야 하며, '상상'은 특권을 가진 사람들의 '잉여'가 아닌 보편적인 '생산적인 활동'으로서 모두에게 평등해야 한다.

이연우
Yonu Lee

전희수
Heesoo Jeon

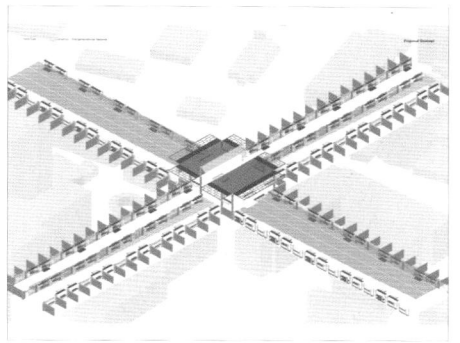

Dongmyo: Intergenerational Network

Tip-Toeing Landscapes

키워드	노인, 청년, 포용성, 시장, 다양성, 지원, 세대교류, 마을, 프로토타입, 활성화
사이트	서울시 종로구 동묘시장 일대
유형	공공시설

키워드	메가스트럭쳐, 유토피아, 미래도시, 공중 정원, 열섬, 쿨링, 인프라, 스케일, 환경적 불평등, 고도, 발전
사이트	서울시 강남구
유형	복합시설

동묘시장은 서울에서 노인들이 주가 되는 장소들 중 여러 세대가 섞이는 거의 유일한 장소이다. 노후화로 인해 많은 재건축이 진행되고 있는 중인 이 곳은 근래 사회적으로 가장 중요한 이슈인 세대 간의 경계를 유연하게 만들 수 있는 잠재력이 보존되어야 한다. 시장 이용자들의 편의를 최대한 유지 또는 더 개선해주는 설계를 통해 시장의 가치를 몰라보는 사람들에게 더 이상 허물어야 할 노후화된 동네로 인식되지 않도록 하는 전략적 개입이 필요하다.

메가스트럭쳐와 원시적 도시가 공존하는 과도기에서 도시인은 주체를 상실한 채 보이지 않는 유토피아를 향해 고개를 처든다. 땅의 감각은 상실하고, 마천루의 꼭대기만을 바라보며 스스로 도시의 부산물이 되기를 자처한다. 거대한 건축물과 인간의 불평등한 관계에 전복이 필요한 지금, 우리에겐 중력을 온전히 느끼며 땅을 디뎌내려는 태도가 필요하다. 허구의 이상향을 향한 무의미한 날갯짓이 아니라 중력을 온전히 감각하는 'Tip-toeing' 형태의 태도를 지향하며, 메가스트럭쳐와 인간이 공존하는 도시를 제안한다.

Dongmyo Market is almost the only place in Seoul where many generations mix among the places where the elderly are the main ones. This place, which is undergoing a lot of reconstruction due to its deterioration, must preserve the potential to make the boundaries between generations flexible, which is the most important social issue these days. Through a design that maintains or further improves the convenience of market users as much as possible, it is necessary to strategically intervene so that it is no longer perceived as an aging neighborhood that needs to be demolished by those who do not know the value of the market.

In the transition period where mega-structure and primitive cities coexist, urbanites raise their heads toward the invisible utopia without losing their subject. The sense of the ground is lost, and he claims to be a by-product of the city by looking only at the top of the skyscraper. Now that we need subversion in the unequal relationship between huge buildings and humans, we need an attitude to step on the ground while fully feeling gravity. It aims for a "tip-toeing" attitude that fully senses gravity, not a meaningless wing movement toward a fictional ideal, and proposes a city where mega-structure and humans coexist.

This project, derived from sudoku, dissolves the rigidity of existing urban planning by creating a city with new movements. The process of searching for blanks in sudoku is a repetitive one of ephemeral movements traversing the borders of the board's axes. By implementing sudoku's ephemeral movements into the city, we're able to observe moments of ephemeral intimacy. Hearing people cheer at bars and cheering along with these people we've never met is a moment of intimacy in which we find joy by blurring our boundaries. This is the ephemeral movement or ephemeral intimacy found in the city; soon to disappear, leaving only a lingering feeling. Densely packing these moments of ephemeral intimacy within our city, we're able to healthily exhaust our feelings, thus creating a vibrant city.

Recovering the fossilized and contaminated land of Samgakji, this project interprets the land as an active entity that creates and helps the life of living things, not the base of the foundation of the building. Pung-Baek(風伯), which is a traditional agricultural method of revitalizing soil, has been applied through structural and morphological methods.

Plowing the soil to breathe and blowing new breath into it, a tree-like structure has been planted on the site. Just as the roots of trees absorb water and nutrients and transport them to leaves through stems, plowed soil rises to the ground and is sprayed on domes and vaults to become new soil for agricultural land. This structure expands the surface area of the soil like a leaf, and a new ecosystem is created, recovering earth by inhaling and exhaling.

김보경
Bogyong Kim

이지윤
Jiyun Lee

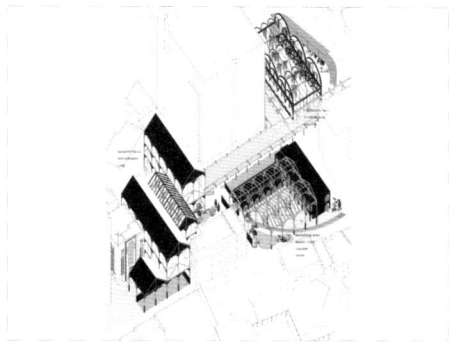

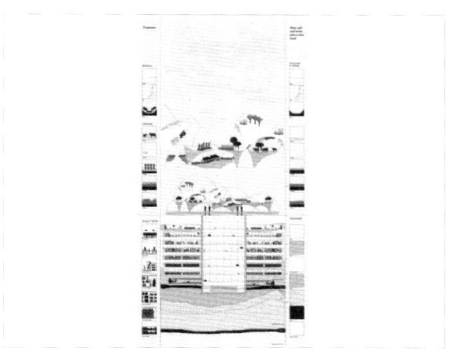

Magnetic Field City

Plowing and Blowing

키워드	스도쿠, 결혼식, 분절, 유대, 아치, 모듈, 관람, 자기장, 회귀, 활기
사이트	서울시 중구 퇴계로, 필동로 일대
유형	상업시설

키워드	화석화, 회복, 토양, 자연, 농사, 생명력, 토목기법, 식생의 다양성, 생태계, 도시체계
사이트	서울시 용산구 한강대로 160 현 한국전력 용산창고 부지
유형	복합시설

기존 도시계획의 경직성을 해소하고 새로운 흐름을 가진 도시계획을 위해 스도쿠놀이에서 시작된 프로젝트이다. 스도쿠에서 빈칸의 숫자를 찾아나가는 과정은 스도쿠판 축의 경계를 가로지르는 일시적인 흐름의 반복이다. 스도쿠의 일시적인 흐름을 도시로 치환하면 도시에서 일시적인 유대감을 느끼는 순간으로 볼 수 있다.

주점에서 누군가의 환호성을 듣고 함께 환호하는 것은 일면식이 없고 다른 위치값을 가진 누군가와 순간의 유대감으로 경계가 흐려지는 즐거움을 느끼는 것이다. 이것이 도시에서 찾은 곧 사라지지만 여운이 남는 일시적인 흐름 즉, 일시적 유대감이다. 이러한 일시적인 유대감이 생기는 순간들이 도시에 밀도있게 채워져 건강한 감정 소모를 하는 활기있는 도시를 만든다.

수없는 건축행위로 오염된 도시의 땅을 회복한다.

땅을 건축물이 들어서는 기초가 아닌 생명을 탄생시키는 능동적 주체로 농법의 관점에서 접근하여 전통 농법 중 흙을 살리는 지혜인 풍백(風伯) 기법을 적용한다.

땅을 숙아 숨을 불어넣고 그 자리에 나무와 같은 건축을 심는다. 뿌리가 흙 속에서 양분을 얻고 기둥을 통해 물과 양분이 이동하여 잎에서 생산활동을 하듯, 숙아낸 땅의 흙은 지상으로 올라와 돔과 볼트 위에 뿌려져 경작지의 흙이 된다. 구조체는 나뭇잎처럼 흙의 표면적을 넓히고 숙아져 올라온 곳에서 새로운 생태계가 만들어진다.

이은후

Eunhoo Lee

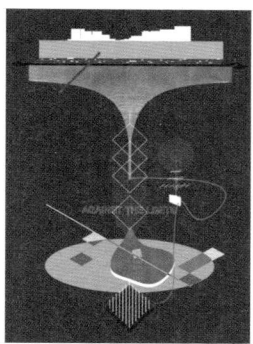

The Fourth Attack: Linking Instabilities

키워드	땅, 기후변화, 홍수, 빗물재사용, 기반시설, 저수조, 펌프, 지형, 일상, 허용한계
사이트	서울시 용산구 한강대로 11길 현 빗물펌프장
유형	복합시설

In 1887, Chief Seattle wrote of his loss in a letter to the white men who occupied his village. We thought the land would stay the same, but it never did. The people who remained left, saying they would undergo another change, and the cycle would repeat itself ad infinitum. In 2022, will our land be forever? We look at Yongsan Mountain at the northern end of the Hangang River Bridge, which is undergoing endless changes due to climate change, sea level rise, and the limits created by urban elements. A greeting to the land waiting for the fourth blow: The Fourth Attack: linking instabilities

1887년, 시애틀 추장은 마을을 점령한 백인들에게 보내는 편지에 상실감을 적었다. 땅은 그대로일 줄 알았지만 결코 그렇지 않았다. 남은 사람들은 또 다른 변화를 겪을 것이며 변화는 무한히 반복될 것이라고 말하며 떠났다. 2022년, 우리의 땅은 영원할까? 기후변화, 해수면 상승, 도시 요소들이 만들어내는 한계가 짙어 끝없는 변화를 겪고 있는 한강대교 북단의 용산을 살핀다. 네 번째 타격을 기다리고 있는 땅에게 보내는 안부인사: The Fourth Attack: linking instabilities

강하림
Harim Kang

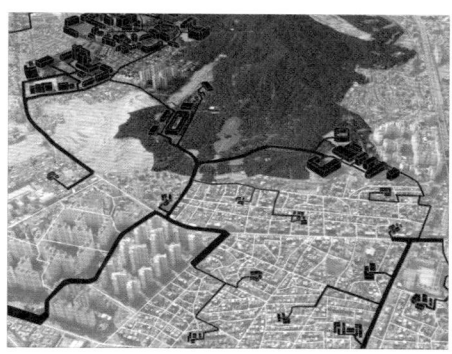

Cul-De-Sac

키워드	도시의 방, 대학, 지식공유, 15분 도시, 탈중심, 용해, 거리, 추출, 중립, 중추
사이트	서울시 동대문구 천장산로4길, 이문로9길 일대
유형	상업시설

Similar typologies of "Jimjilbang", among the rooms in Korea, which has a neutral living room space on the center, surrounded by various steam rooms, are extracted from the urban fabric. Their neutral space has the character of staying rather than passby, so people take a rest while looking around and create communities. Among them, dead end and adjacent houses are observed and redefined as knowledge sharing startup labs for universities in Cheonjang Mountain. Universities get urbanized and expanded, blurring the boundaries of campuses, suggesting new urban rooms to practice, research, create and share regardless of universities or major.

한국의 방들 중 중앙은 거실의 중립적 공간이 있고, 주변은 찜질의 프로그램 안에서 다양한 분위기의 방들로 둘러싼 찜질방의 위상과 유사한 도시 조직을 추출한다. 이들이 가지는 중립적 공간은 이동보다는 머무름의 성격을 띠어 사람들이 주변을 보며 휴식하고 커뮤니티가 발생한다. 그 중 막다른 길과 인접한 주거들을 관찰하며 천장산 자락의 여러 분야의 대학들의 지식공유연구소로 재정의한다. 대학들이 점점 도시화되고 확장되어 캠퍼스의 경계가 흐려지며, 대학과 전공에 상관없이 연습하고, 연구하고, 창작하고, 나누는 새로운 도시의 방들을 제안한다.

파트 3.
Part 3.

1. 서문: 졸업에 대한 생각과 소회
2. 메인 드로잉
3. 목차
4. 리서치 작업
5. 사이트와 프로그램
6. 매스 스터디
7. 배치도
8. 평면과 단면, 투시 드로잉
9. 기술 스튜디오
10. 전시 기록

1. Preface
2. Main Drawing
3. Process
4. Research Book
5. Site + Program
6. Mass Study
7. Site Plan
8. Plan + Section + Perspective Drawing
9. Techinical Studio
10. Exhibition Map + Model

While each studio and each student's project is slightly different, the overall design process is similar, so I've organized it into six steps.

In part 1, I wanted to start with an article that summarizes the big picture of my work and share the big picture of graduation and design work.

Part 2 starts the portfolio with the main drawing of the graduation project, which is a short introduction to the big picture of design.

In part 3, you can read how the students worked on their designs. Not everyone followed the same process, so this part of the portfolio provides a more detailed look at the design process as it was woven together through editing.

In parts 4–6, I've organized the tasks and processes that led to the drawings and perspective drawings into common steps, although some students may not have used certain parts.

Parts 7 and 8 contain the results of the drawn work, which, when opened, are much smaller than the size of the actual drawings, and cannot be fully captured on the ground, so it is regrettable that the scale and actual impression felt in the exhibition are not fully conveyed.

Part 9 contains the technical detail drawings that the students needed for the work they did in the Technology Studio class.

Finally, Part 10 is a record of the graduation exhibition, including the layout of the exhibition hall, the types of models used in the exhibition, how they were made, photos of the models, and the graduation helpers.

스튜디오별, 학생별 프로젝트가 진행되는 방식이 조금씩 상이하지만, 큰 맥락에서 진행되는 설계의 방식은 비슷하기에 6가지의 단계로 나누어 엮었다.

1번 파트를 통해 본인이 생각하는 작업의 큰 맥락을 정리하기도 하고, 졸업과 설계 작업에 대한 큰 그림을 나누는 글로 시작했으면 했다.

2번 파트를 통해 졸업 작업의 메인이 되는 드로잉으로 설계의 큰 줄기를 짚고 넘어갈 수 있는 단초로 포트폴리오의 설명이 시작된다.

3번을 통해 학생들의 설계 작업 진행 방식을 읽을 수 있다. 모두가 같은 프로세스로 진행한 것이 아니기에, 편집을 통해 엮은 설계의 과정들은 이 파트로 좀 더 자세히 들여다 볼 수 있다.

4-6번 파트를 통해 도면과 투시 드로잉의 결과가 나오기까지의 작업들과 프로세스들을 공통적인 단계들로 묶었고, 학생들에 따라 특정 파트가 없는 경우도 존재한다.

7-8번 파트를 통해 도면화된 작업의 결과물을 담았는데, 책을 펼쳤을 때 실제 도면의 사이즈보다 훨씬 작기에 지면에 충분히 담아낼 수 없어 전시장에서 느껴졌던 스케일과 실제의 감동이 온전히 전달되지 않는 아쉬움이 있다.

9번 파트를 통해 학생들이 기술스튜디오 수업에서 다룬 작업에 필요한 기술적인 디테일 도면을 담았다.

마지막으로 10번 파트에서는 졸업 전시에 대한 기록으로, 전시장 배치 형태와 전시에 사용된 모델의 종류, 만든 방식과 모델 사진, 졸업 도우미 등 전시에 관련된 기록으로 마무리지었다.

Cul-De-Sac Calvin Chua
 Studio

강하림
Harim Kang

Since the "Back to the Earth" exhibition was not a crossover of projects, but rather a theme that emerged from a common event or situation for all of us as we approached graduation, we created display stands to unify the exhibition. A display table with the studio's theme, description, and name hung next to each other separated us from our neighbors while also breaking up the beginning of each exhibition. The exhibit I organized is characterized by categorizing and separating parts that are meant to be viewed over a long period of time. I wanted people to be able to choose their experience based on their level of interest in the project. You could dive into the portfolio, observe the mockups, or skim over the minimal descriptions and images. But more importantly, I wanted the two main models to be viewed from multiple angles, and to be arranged in a way that takes into account their relative positions in the real world in order to show their connections and sequences.

Graduating with this project was a bit of a disappointment. In fact, I didn't "graduate" at all. There were three types of buildings that I wanted to plan, but I was only able to finish two of them. I even imagined the third type as a vessel for the main function of the program. This was especially disappointing because I felt that the synergy between the three types would be stronger if they were placed together. If you look at graduation as a journey, it's been a five-year journey that has been filled with regular cycles of design studio days that have left me with motion sickness, but it's also been a lot of fun just spending time with my peers. I learned that anything can be completed if you don't let go, and I realized what I was interested in by looking at my portfolio, which was consistently looking in the same direction. The final week of preparation for the exhibition was the most intense experience. It was a great learning experience.

"Back to the Earth" 전시는 프로젝트들에서의 교집합이 아닌, 졸업을 앞둔 우리들에게 놓여있던 공통의 사건 혹은 상황에서 나타난 주제였기 때문에 전시의 통일성을 위해 전시대를 제작했다. 스튜디오 주제 및 설명, 이름이 나란히 걸린 전시대로 옆사람과의 칸막이면서 동시에 각 전시 관람의 시작을 끊어주었다. 내가 구성한 전시의 특징은 장시간 머무르며 관람할 항목을 분류 및 분리하여 배치한 것이다. 프로젝트의 관심도에 따라 관람방식을 선택할 수 있으면 좋겠다고 생각했다. 포트폴리오를 자세히 들여다 볼 수도 있고, 모형을 관찰할 수도 있고, 최소한의 설명 및 이미지만 슥 보고 지나칠 수도 있다. 하지만 이것보다 더 신경쓴 부분은 주 모형 2개를 다각도에서 관람가능하며, 서로의 동선에서의 연결 및 시퀀스 표현을 위해 실제의 상대적 위치를 고려한 배치이다.

이 프로젝트로 졸업하며 아쉬움이 많았다. 사실은, "졸업"하지 못하였기 때문이다. 계획하고픈 건축유형이 3가지였으나, 2가지만 겨우 끝낼 수 있었다. 심지어 세 번째 유형이 프로그램의 주기능을 담는 그릇으로 상상했었는데 말이다. 이들이 함께 놓여있어야 시너지가 생기며 논리의 설득력이 강해진다고 느끼기 때문에 더욱 아쉬웠다. 졸업을 하나의 여정으로 본다면 일정한 주기로 꾸준히 다가오는 설계날로 멀미가 일 듯이 고되기도 했지만 동료들과 함께 시간을 보낸 것만으로 그만큼 즐거운 5년이었다. 놓지만 않는다면 무엇이든 완성된다는 것을 배웠고, 내가 어떤 주제에 관심있는지를 꾸준하게 같은 방향을 바라보고 있는 그동안의 포트폴리오들을 보며 깨달았다. 전시를 준비하는 마지막 관문의 일주일은 경험의 강도가 가장 짙었다. 정말 큰 공부가 되었다.

SITE PROPOSAL_
EXPANSION/ DECENTRALIZATION OF
UNIVERSITIES THROUGH NEUTRAL SPACES

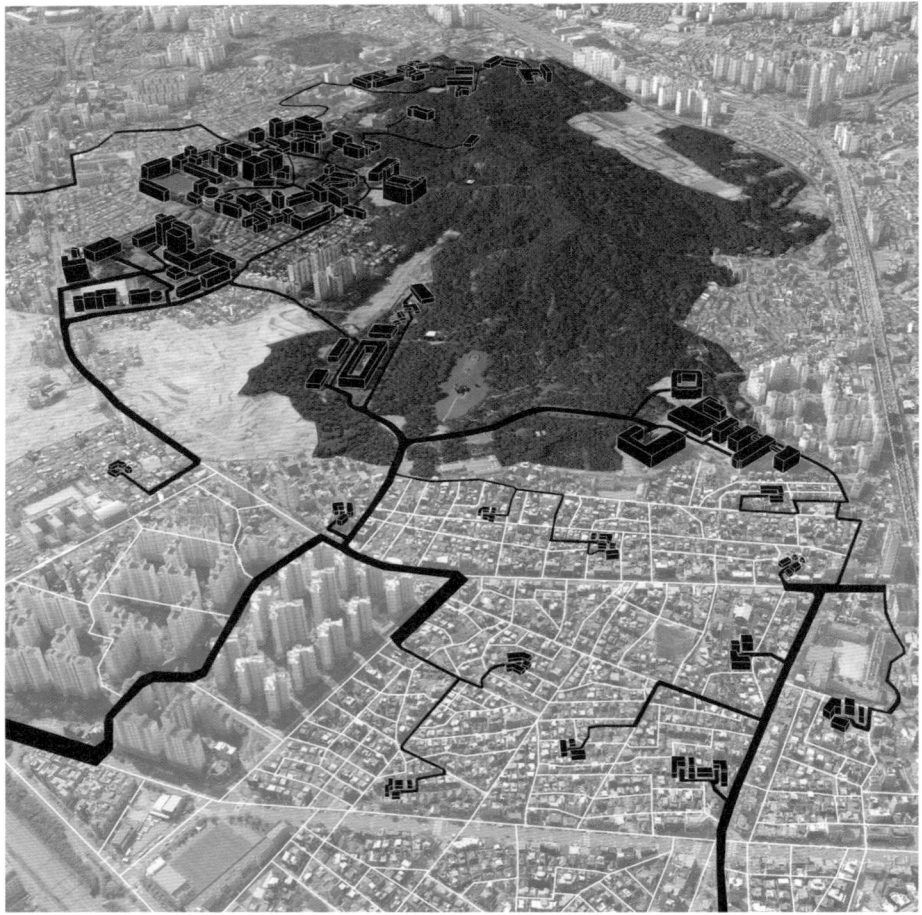

The project is titled "Cul-de-sac_ Neutral space as a Typology of Knowledge Sharing Lab". I defined a neutral space as a type of city and extract these organizations. Then I architecturalize the extracted neutral space. The studio's core concern is to rethink the approach to the 15-minute city from the perspective of strategic reinforcement of existing neighborhoods in Seoul through 'urban rooms'. Between the micro and macro scales of the existing city rooms, we propose a new 'city room' that benefits 1000 local residents.

Korean educational institutions, especially schools in Seoul, are located at the foot of mountains. Since ancient times, it has been believed that academic learning is more effective when one is close to nature and away from the world. There's also a reason why land near mountains is more economically advantageous than flat land. Therefore, schools were built in a circle around mountains.

The development of industrial technology and declining fertility rates have led to an aging population and an imbalance between young and old, and the development of information and communication technology has reduced the need for large campuses, especially those of universities. The main role of universities has shifted from providing higher education to being a springboard to society by providing problems and opportunities, which means that there is a need for communication among diverse students and a place where they can gather anytime and anywhere. Therefore, universities are increasingly urbanizing and expanding, blurring the boundaries of the campus, and proposing new urban rooms to practice, research, create, and share knowledge regardless of university and major.

프로젝트 제목은 "Cul-de-sac_ Neutral space as a Typology of Knowledge Sharing Lab"이다. 도시의 한 유형으로 중립적 공간(neutral space)를 정의하였고, 이러한 조직을 추출한다. 추출한 중립적 공간을 건축화한다. 스튜디오의 핵심 관심사는 '도시의 방'을 통한 서울 기존 지역의 전략적 강화의 관점에서 15분 도시에 대한 접근 방식을 재고하는 것이다. 기존의 도시의 방이 가지는 미시와 거시의 두 스케일 사이에서, 1000명의 지역주민에게 혜택을 주는 새로운 '도시의 방'을 제안한다.

한국의 교육기관, 특히 서울의 학교는 산자락에 위치해 있다. 예부터 학문의 배움이라 함은 자연과 벗삼아 속세와 멀리 떨어져야 더 효과적으로 이루어진다고 보았다. 산 인근의 땅이 평지보다 경제적 면에서 유리한 이유도 있다. 따라서 학교들은 산 주위를 빙 둘러 위치하는 형상을 띤다.

산업기술의 발달 및 출산율 감소로 고령화와 청년-노년층의 인구불균형이 발생하며, 정보통신기술의 발달로 큰 캠퍼스, 특히 대학의 거대한 캠퍼스의 필요성이 감소하고 있다. 대학의 주 역할이 고등교육 제공에서 문제 및 기회제공을 통한 사회로의 발판으로 이동하였으며, 이는 다양한 학생 간의 소통과 언제 어디서든 모일 수 있는 장소가 필요함을 의미한다. 따라서 대학들이 점점 도시화되고 확장되어 캠퍼스의 경계가 흐려지며, 대학과 전공에 상관없이 연습하고, 연구하고, 창작하고, 지식을 나누는 새로운 도시의 방들을 제안한다.

1. 15분 도시 분석
2. 도시의 방 분석
3. 사이트 관찰
4. 중립적 공간 정의 및 추출
5. 큰 스케일에서의 제안, 작은 스케일에서의 건축화

The 15-minute city is a concept of a self-sufficient urban village within a 15-minute distance (1 kilometer walking distance). I looked at Clarence Perry's neighborhood unit and learned that schools are the backbone of a city, and that cul-de-sacs are a type of design to prevent through traffic from causing accidents and urban fragmentation.

In the urban room, I looked at Korean visitation, where rooms from different environments are juxtaposed in a jjimjilbang, and a neutral space is defined in the middle living space that encompasses them. In this way, the urban organization was extracted by looking at the site through the lens of the interpretation of the 15-minute city and the city room.

The site is a mountain surrounded by universities and research institutes that educate various disciplines in the humanities, social sciences, sciences, and arts. Proposals were made at the urban scale for the connection of the university and its expansion into the city, and at the architectural scale to build the extracted urban organization, especially cul-de-sac, while maintaining the potential of the space.

15분 도시는 15분 거리(도보거리 1km)에서 자급적 기능을 수행하는 도시의 마을 개념이다. 그 중 Clarence Perry의 근린주구론을 살펴보았으며, 학교가 중추가 되는 도시이고, 사고와 도시의 분절을 발생시키는 통과교통을 막기 위한 쿨데삭(막다른 길) 디자인이 한 유형으로 작용함을 배웠다.

도시의 방에서는 한국의 방문화를 살펴보았고, 그 중 찜질방의 여러 환경의 방들이 병치되고 그것들을 아우르는 중간의 거실공간에서 중립적 공간의 정의가 이루어졌다. 이렇게 15분 도시와 도시의 방을 해석한 렌즈로 사이트를 들여다보며 도시 조직을 추출하였다.

사이트는 인문, 사회, 과학, 예체능의 다양한 학문을 교육하는 대학 및 연구기관이 산을 주위로 둘러싼 형태이다. 대학의 연결 및 도시로의 확장을 위한 도시적 스케일에서의 제안을 하고, 추출된 도시 조직, 그중에서도 쿨데삭을 공간의 잠재성을 유지하며 건축화하는 건축적 스케일에서의 제안을 진행하였다.

Cul-De-Sac

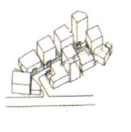

1. dead end as a neutral space

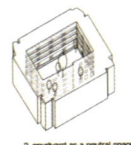

2. courtyard as a neutral space

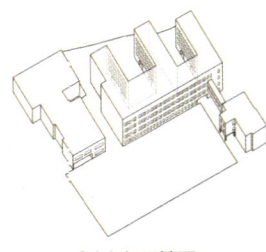

3. schoolyard as a neutral space

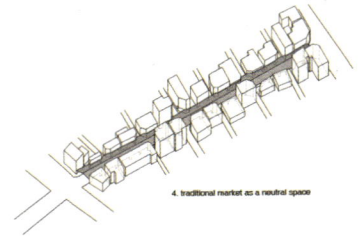

4. traditional market as a neutral space

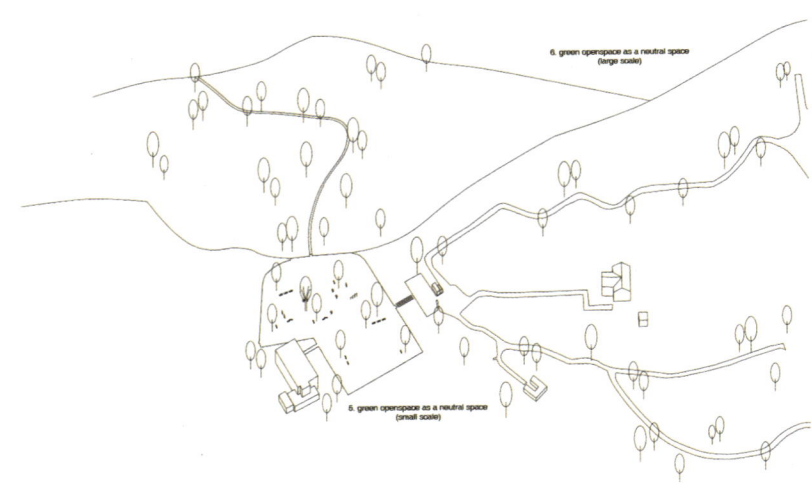

6. green openspace as a neutral space (large scale)

5. green openspace as a neutral space (small scale)

The central living space of the jjimjilbang is defined as a neutral space. Neutral spaces are characterized by lingering rather than moving, where people relax and community occurs around them. They are also spaces that are defined by the context and boundaries of their surroundings, and their characteristics are weaker than those of the periphery. As we explored the site, we found several types of such spaces.

1) Dead-end streets (cul-de-sac)
2) Middle courtyard
3) Schoolyard
4) Markets
5) Small-scale open space with exercise equipment
6) A large-scale open space with multiple paths leading to it.

These neutral spaces are characterized by the fact that they are branching points before they lead to different houses, rooms, classrooms, shops, sports equipment, and trails, and they are characterized by being still because they are the end of the public space. The paths that diverge have different destinations, environments, and atmospheres, and this preceding neutral space becomes a place where people with different choices and purposes can gather and mix. People can stay here for a long time comfortably because they have their next destination behind them. Therefore, this neutral space has the potential to function well, both physically and psychologically.

찜질방의 중앙 거실공간을 중립적 공간(neutral space)로 정의하였다. 중립적 공간은 이동보다는 머무름의 성격을 띠어 사람들이 주변을 둘러보며 휴식하고 커뮤니티가 발생한다. 또한 주변의 컨텍스트와 경계에 의해 정의되는 공간이며 그 특성이 주변부보다는 약한 공간을 의미한다. 사이트를 답사하며 이러한 성격을 띠는 몇 가지 유형을 발견하였다.

1) 막다른 길(쿨데삭)
2) 중정
3) 학교 운동장
4) 시장
5) 운동기구가 놓인 작은 스케일에서의 오픈 스페이스
6) 여러 갈래의 산책로로 이어지는 큰 스케일에서의 오픈 스페이스

이러한 중립적 공간은 서로 다른 집, 방, 교실, 가게, 운동기구, 산책로로 이어지기 전의 분기점이라는 특징이 있으며, 형태적으로는 공적공간의 말단이기 때문에 고여있다는 특징을 가진다. 그렇게 갈라진 길은 다 다른 목적지와 환경, 분위기를 가지고 있으며 그 전단계인 이 중립적 공간은 서로 다른 선택과 목적을 가진 사람들이 모이고 쉴 수 있는 장이 되는 것이다. 이 곳의 사람들은 다음 목적지를 뒤에 두고 있기 때문에 오히려 편안하게 오래 머물 수 있다. 그렇기에 이 중립적 공간은 물리적 형태로나 사람의 심리적인 면에서나 잘 기능될 가능성이 있는 공간인 것이다.

Cul-De-Sac

SITE OBSERVATION_
DEFINITION OF NEUTRAL SPACE

dead end road as a neutral space

Dead end road is a space with possibilites of communities to occur. Neighbourhoods share their lives, events, hardships and events. Each gates shows the character of the house. Childrens play, mothers do chores and chat. What if dead ends become to an architecture? How can the space be used?

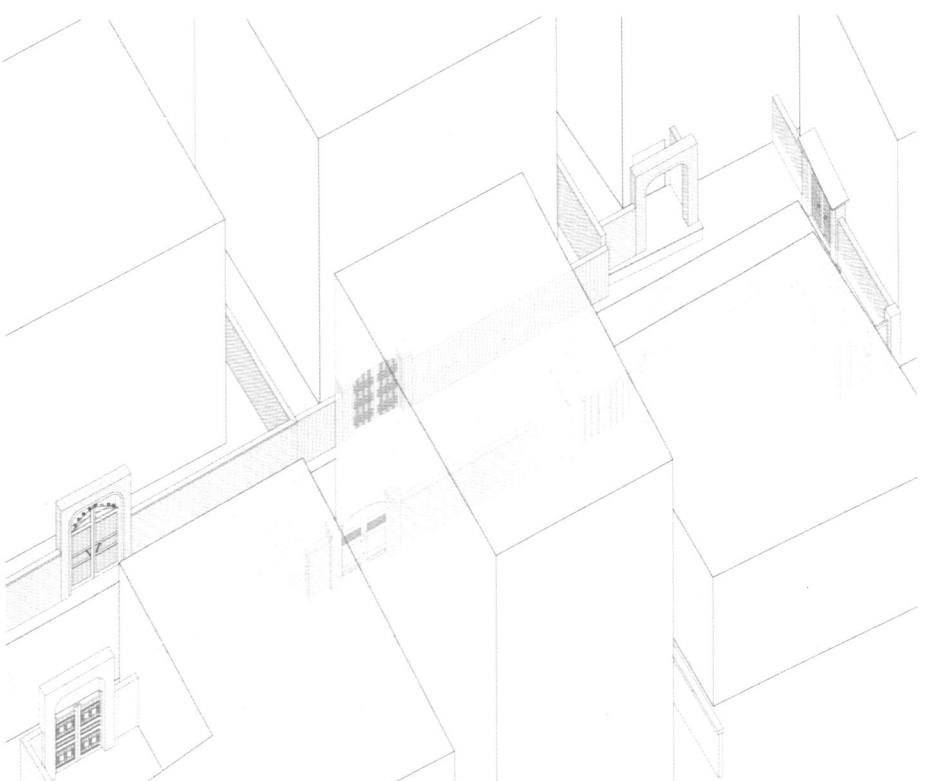

The most program- and content-free of all, yet ubiquitous in the city, dead-end streets were chosen as the type of neutral space to be built. I felt that it had the most potential to be used differently depending on its morphological differences such as width, depth, and slope. A space where people can grow bean sprouts, chop garlic, hang laundry, share food, play, watch movies, assemble furniture, and grow plants.

그 중 가장 프로그램 및 내용이 없는, 그렇지만 도시 어디에나 존재하는 막다른 길을 건축화할 중립적 공간의 유형으로 선정하였다. 너비, 깊이, 기울기 등 형태적 차이에 따라 다르게 사용될 수 있는 가장 잠재성이 큰 공간으로 느꼈다. 함께 콩나물, 마늘을 다듬고, 빨래를 널고, 음식을 나누고, 무궁화꽃이 피었습니다 놀이하고, 영화를 보고, 가구를 조립하고, 식물을 가꿀 수 있는 공간.

Patrick Abercrombie's Potato Plan and The Situationist Plan were used to observe the site. The Potato Plan is an internal view of the city as a collection, a territory, and the Situationist Pan is an external view of the city as a situation, a connection, a sequence.

The site consists of six universities and research facilities that share Mount Cheonjangsan. University campuses are places that people who don't belong to them rarely visit, even if they live in the area for years, or people who do belong to them stay in a particular place for years, so they are often visited by chance or under special circumstances. Imagine the experience of walking through a series of universities along back roads, rather than adjacent to the main road where apartment redevelopment is taking place, and converging at certain points.

Patrick Abercrombie의 Potato Plan과 The Situationist Plan으로 사이트를 관찰하였다. Potato Plan은 도시를 집합, 영역으로 보는 방식으로 내부적 요소를 들여다보는 관점이며, Situationist Pan은 도시를 상황, 연결, 시퀀스로 보는 방식으로 외부적 요소를 찾아보는 관점이다.

사이트는 천장산을 공유하는 6개의 대학 및 연구시설로 이루어져 있다. 대학 캠퍼스는 소속되지 않는 사람의 경우 몇 년을 그 지역에 거주해도 잘 가지 않거나, 소속된 사람의 경우 몇 년을 살아도 특정 장소에만 머무르는 곳이라 우연한 계기나 특수한 상황에 무심결에 방문하는 장소이기도 하다. 아파트 재개발이 이루어지는 주 도로 인접지보다 뒷길을 따라 대학들을 이어 발길이 닿는대로 걷다 어떤 지점들로 고이는 경험을 상상해본다.

Cul-De-Sac

THE SITUATIONIST CITY

student residents:
A_ 48-44-40-41 55
B_ 37-29 34 36
C_ 42-46-52
D_ 53-54 45-44-40 6
E_ 24 26 27 28
F_ 19-22 16 20

residents:
A_ 56-58 49
B_ 1 2 3-5 16
C_ 15 21
D_ 45 47 50 51 57-58
E_ 11-13-14-46 7 12
F_ 23 28 33 36

students outside the neighbourhood:
A_ 32-31-30
B_ 52-53

On the site, the locations for the concrete plan are represented by a one kilometer belt. Neutral spaces, dead ends (culdesac), become eventualities that are encountered while walking through the streets, some of which are transformed into architecture. The belt connects universities and becomes a spine where knowledge and experience collide to produce ideas. Among them, the 1.5km belt connecting Korea National University of Arts to Kyung Hee University was selected as the final site and proposed as a startup space for 1000 people. It is a smaller-scale proposal that can be more closely connected to actual urban organizations than existing coworking spaces.

사이트에서 구체적인 계획을 위한 위치들을 1km의 벨트로 표현하였다. 중립적 공간인 막다른 길(쿨데삭)은 거리를 거닐면서 마주하는 이벤트적 요소가 되고, 그 중 일부는 건축으로 변모된다. 벨트는 대학들을 연결하며 지식과 경험이 충돌하며 아이디어를 생산하는 하나의 척추가 된다. 그 중 한국예술종합학교에서 경희대를 잇는 1.5km 벨트를 최종 사이트로 선정하여 1000명을 위한 스타트업 공간으로 제안한다. 기존에 존재하는 공유오피스 사례보다 실재하는 도시 조직과 더욱 긴밀히 연결될 수 있는 보다 작은 스케일의 제안이다.

REGENERATION PROPOSAL_
STRATEGY CONCEPT

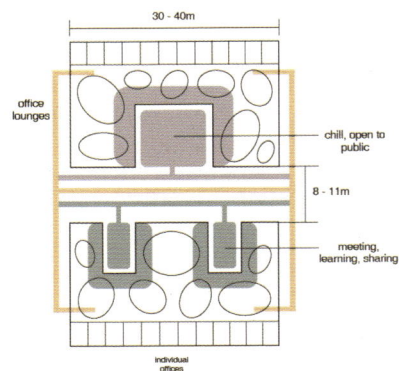

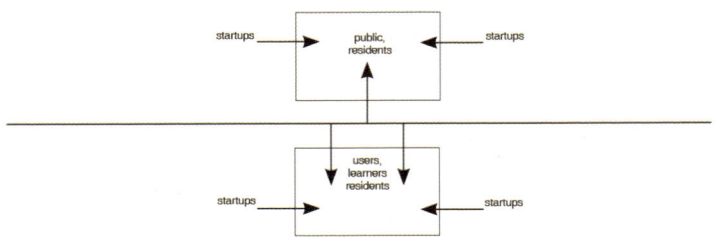

The conceptual idea for the space plan was to think about the needs and programs of startup groups, students, and local residents, and to divide the space into three categories: space for creators, space for users and students, and space for the public and local residents, so that these groups can relate but not mix. The three approaches to this are a growing linear strip, a diamond hall, and two facing masses.

The linear strip uses surplus space and dead-end streets that used to be used for parking next to the building to create a linear, minimal office space that will grow over time, while the diamond hall is for more formal and larger events for students and locals. The two facing masses are responsible for the startup's main programs and face the main road of the 1.5 km belt, a proposal that shows pedestrians the scene of knowledge sharing.

공간계획에서의 컨셉 아이디어는 스타트업 그룹, 학생들, 지역 주민들에게 필요나 프로그램을 생각해 창작자를 위한 공간, 사용자와 학생들을 위한 공간, 공공과 지역주민을 위한 공간의 세 카테고리를 나누어 이 그룹들이 관계를 맺되 쉬이지 않도록 생각하였다. 이를 위한 세 가지 접근은 자라는 선형 스트립, 사각형 홀, 그리고 마주보는 두 매스이다.

선형 스트립은 건물 옆 주차로 사용되던 잉여공간과 막다른 길을 사용하여 선형의 최소한의 오피스 공간을 만들어 시간이 지남에 따라 성장하게 하였고, 사각형 홀은 학생들과 지역주민들을 위한 보다 공식적이고 큰 이벤트를 위한 홀이다. 마주보는 두 매스는 스타트업의 주 프로그램을 담당하며 1.5km 벨트의 주 도로를 사이로 마주보아 보행자에게 지식 공유의 현장을 보여주는 제안이다.

REGENERATION PROPOSAL_
THREE DIFFERENT APPROACHES

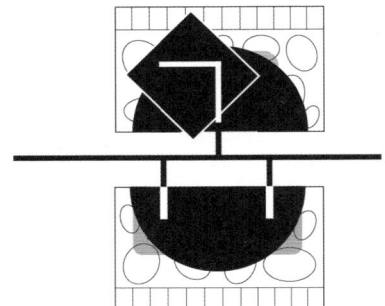

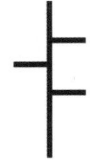      

GROWING STRIP

Growing strip is a strategy of adjusting the co-working space into a smaller scale that interacts actively with the existing blocks. Original dead ends become community space for both residents and users.

DIAMOND HALL

Diamond hall is an event hall that gathers people. Original dead end becomes the main entrance to the building and opens inward.

TWO MASS FACING

Two mass are facing inbetween the road, showing passengers the live creating scene. dead end is made for learning and community space.

In the linear strip, a 3-meter-wide road next to the building was used, and the existing dead-end street was proposed as a local hall where community could occur. The strip connects Korea National University of Arts, Hankuk University of Foreign Studies, and Kyung Hee University and encourages students to meet and interact naturally. The ground floors of the existing buildings were reprogrammed to include office lounges, gardens, cafes, and creative spaces for residents and students. Over time, the strip can expand and grow into a market-like center.

The diamond hall is where events such as forums, seminars, and celebrations take place. An existing dead-end street was utilized as the main entrance to the building, leading to a circular staircase and culminating in a single point. There are minimal 3-meter-wide staging rooms in the corners, and the different activities in these rooms are visible as multi-colored windows when viewed from inside the hall.

선형 스트립에서는 3m 너비의 건물 옆 도로를 사용했고, 기존 막다른 길은 커뮤니티가 발생할 수 있는 지역 홀로 제안하였다. 이 스트립은 한예종, 외대, 경희대를 모두 잇고, 학생들의 자연스런 만남과 상호작용을 유도한다. 기존 건물들의 1층을 프로그램만 바꾸어 주민들과 학생들을 위한 오피스 라운지, 정원, 카페, 창작공간으로 구성하였다. 시간이 지나며 이 스트립이 확장되어 마치 시장과 같은 하나의 센터로 성장할 수 있다.

사각형 홀은 포럼, 세미나, 축전 등의 이벤트가 일어나는 홀이다. 기존의 막다른 길을 살려 건물의 주 진입로로 설정해 원형 계단으로 이어지며 하나의 점으로 귀결되고자 하였다. 모서리에 3m 너비의 최소한의 준비 공간들이 있고 이 방들의 서로 다른 행위들이 홀 내부에서 바라볼 때 여러 색의 창들처럼 한눈에 보여진다.

The form of the proposed masses is not distinctive to the site, but rather universal that can be inserted into another site with similar urban organization and phenomena.

There are five sites, one in each 300-meter radius of the 1.5-kilometer belt. Linear strips are located at both ends of the belt, acting as a multiplication and extension of the surplus space that can function temporarily into a useful space.

A diamond hall is placed in the center of the belt, two masses that are facing each other, which contain the actual function, are facing the main road. The linear strip and the diamond hall use the existing cul-de-sac organization to emphasize its character, while the two facing masses borrow from it by inserting spaces that act as architectural dead ends.

제안한 매스들의 형태는 사이트의 특수성을 담고 있기보다 비슷한 도시 조직과 현상을 가진 또 다른 사이트에 삽입될 수 있는 보편성을 가진다.

1.5km 벨트의 300m 반경에 1개씩 총 5곳에 배치하였다. 선형 스트립은 일시적으로 기능할 수 있는 잉여공간을 효용있는 공간으로 증식 및 확장하는 역할을 하여 벨트 양단에 위치한다.

사각형 홀은 벨트의 중심, 마주보는 두 매스는 실질적 기능을 담는 곳으로 주 도로를 사이로 마주보게 배치한다. 선형 스트립 및 사각형 홀은 기존 쿨데삭 조직을 사용하여 그 특성을 강조하였고, 마주보는 두 매스는 건축에 막다른 길의 역할을 하는 공간을 삽입하여 그 특성을 차용하였다.

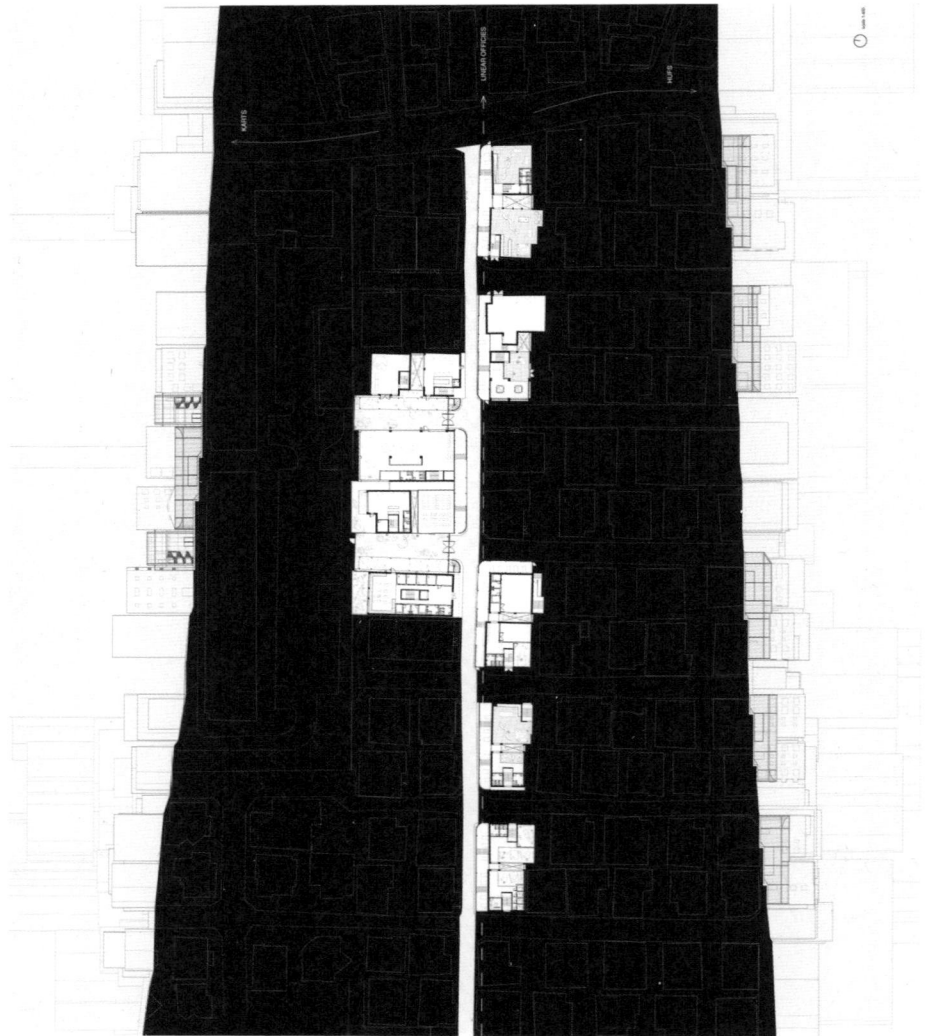

Plan + Section   Harim Kang

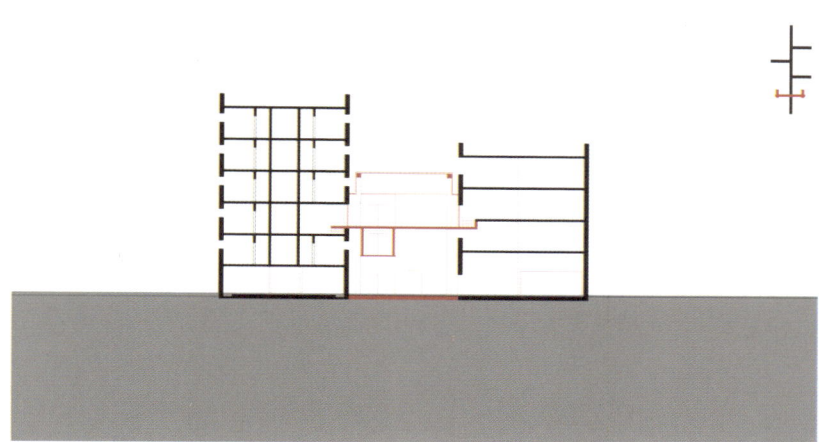

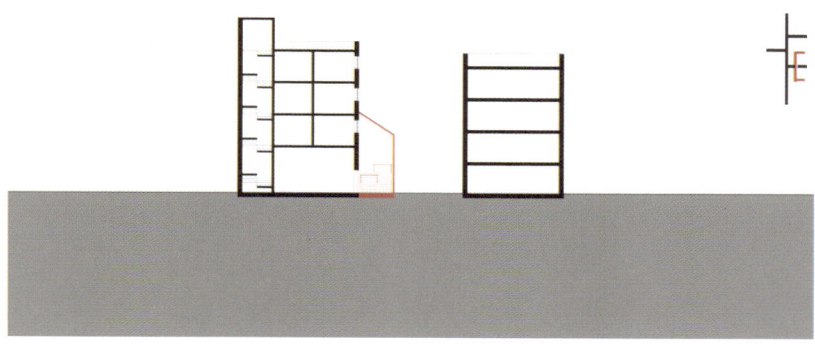

Cul-De-Sac

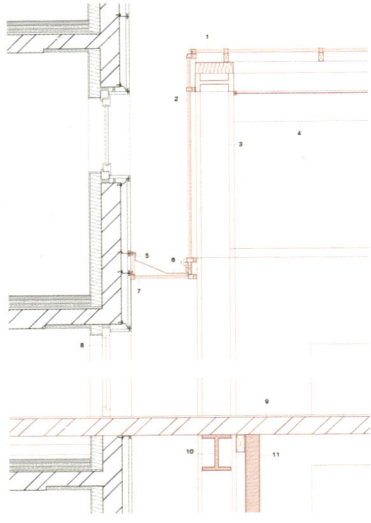

Facade section   scale 1:80

1. double glazing :
   8mm lam. safety glass + 16mm cavity + 10mm lam. safety glass
   100/50mm aluminium joint bar
   120mm insulation
2. double glazing :
   8mm lam. safety glass + 16mm cavity + 10mm lam. safety glass
3. 300/200/40/40mm I-steel arch curved beam
4. interior awning rail+ sheets
5. cladding for stone slate,
   25mm spacing aluminum bar attatched to 20mm spacing wood
6. steel gutter for rain drainage
7. double glazing :
   8mm lam. safety glass + 16mm cavity + 10mm lam. safety glass
8. 2100/750mm double acting swing door
9. floor construction:
   20mm epoxy coating
   200mm concrete slab
10. 300/400/40/40 I-steel beam
    400/400/40/40 I-steel column
11. 150/100mm steel column hung by 100/200mm aluminum joint bar

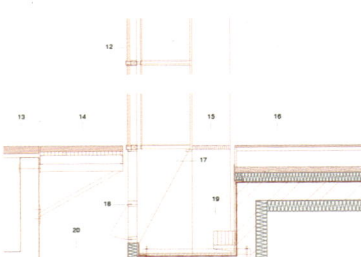

Foundation section   scale 1:80

12. 120/50mm aluminium mullion for double glazing
    36/300mm bracing glass fin
13. 80mm sandstone on 50mm grit bed
14. cover fresh air intake shaft:
    60mm sandstone panel with ventilation slits
    40mm grid
    steel I-section bracket 100mm deep
15. 40mm ventilation slits
16. floor construction:
    40mm sand-lime stone slabs
    40mm tile cladding for access floor
    180mm cavity for access systems
    60mm aerated concrete
    100mm insulation
    PE-film separating layer
    200mm reinforced concrete
    120mm insulation
17. steel bracket as bearing for glass fins
18. aluminium air inlet flap
19. convector
20. drainage with gravel

The atrium, which had an existing dead end in the linear strip, is attached to the neighboring building with a transparent shell for visibility and sunlight inflow, so shading and air conditioning are important. Therefore, exterior and interior shades were installed on the roof, and heat recovery ventilation was carried out from the first floor to the roof.

In addition, a part of the floor and wall is planned as an access floor and wall as various activities are potential.

In terms of architectural details, rainwater and ventilation are considered. There is a path and drain for rainwater to flow, and an outside air inlet is installed near the foundation so that the indoor air could be taken out to the floor.

선형 스트립에서 기존 막다른 길이 있었던 아트리움은 가시성 및 주변 건물의 햇빛유입을 위하여 투명한 외피를 가지고 인접건물과 붙어 있으므로 차양 및 공조가 중요하다. 따라서 지붕에 외부차양 및 실내차양을 설치하고 1층 바닥에서 지붕으로 열회수환기가 이루어지게 하였다.

또한 다양한 활동이 잠재되어 있으니 바닥 및 벽의 일부를 엑세스플로어 및 벽으로 구획하였다.

건축의 디테일 면에서는 우수와 환기에 대해 생각하였다. 우수가 흐르는 길 및 드레인을 만들고 기초부근에 외기 도입구를 설치하여 바닥으로 취출되도록 하였다.

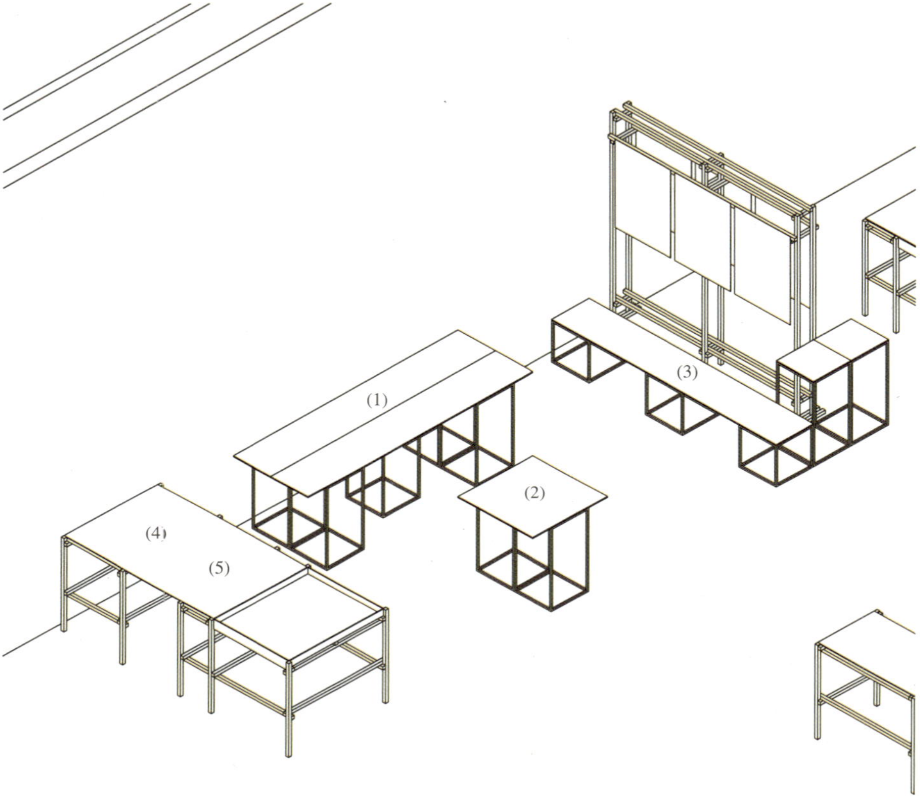

Cul-De-Sac

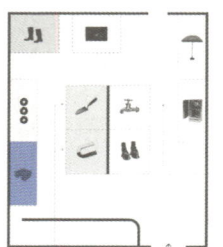

1. Main Model (1)
2. Main Model (2)
3. Structure Model
4. Site Model (1)
5. Site Model (2)

졸업도우미　이혜정
　　　　　　이원준. 이수. 이명재. 윤재원.

MAIN MODEL

Scale 1:100
Box Paper + Royal Board, 80 by 240 cm, 60 by 80 cm

SITE MODEL

Scale 1:2000
Paper & 3D print, 60 by 120 cm

STRUCTURE MODEL

Scale 1:100
Paper & Woodrock, 45 by 90 cm

Magnetic Field City	Seungmo Seo Studio
	김보경 Bogyong Kim

Preface — Bogyong Kim

Before I started my graduation project, I was reminded of the projects that I always felt disappointed with. I knew what made them disappointing without having to go through each project. Slightly bland stories, unfinished and open ends, and time-crunched designs all as a result of inexperience, which ultimately clouded the intensity of each year. Unsure that this year would be any different, I went into this years project. I felt uneasy; skeptical that my graduation project would be any different than before, all while hoping I wouldn't be ashamed of my graduation. Throughout the year and up to the very end, I deeply experienced how grueling graduating really was. Being physically and mentally fatigued all year, I found myself contemplating if I should just give up and graduate the following year, as probably many other graduating students have. Nonetheless, with ample support from those around me, I found myself at the finish line, body and soul intact. I hope to express that I haven't forgotten the abundant things and people I've been grateful for over the past year.

I wanted this year's work to be a little different from my past work, which always seemed slightly aloof. The decorative stories, complex language, and eccentric demeanor all felt redundant. Moving away from concerns of the distant future or the distant past, I wanted the present and my own emotions to find themselves within my work. I wanted my work to be something that many people could relate to, and to be a space that you'd truly want to find yourself in. I spent a year asking myself if I was getting closer to my goal, and there were times I was convinced so, but moments of regret and doubt would always come about. I believe I found something of an answer after the exhibition was over, through a few words written in my visitors log: "I could only wish for the brilliant and eternal memory this would be, to take place in a space as such." I am thankful to be able to conclude my work on these words from an anonymous visitor.

졸업 프로젝트를 시작하기에 앞서 언제나 아쉬운 마음이 들었던 프로젝트들이 떠올랐다. 무엇이 아쉬운 마음을 만들었는지 사실 그 전 작업들을 열어보지 않아도 알고있었다. 부족한 경험이 만든 살짝 밋밋한 이야기와 매듭 짓지 못한 이야기 그리고 시간에 쫓긴 디자인은 매년 치열했던 시간들을 흐리게 만들었다. 올해는 그렇지 않을 것이라 장담할 수 없는 상태와 마음가짐으로 프로젝트를 시작하였다. 졸업작품이라고 다를 수 있을 것인가 하는 의심과 그래도 부끄럽지 않은 졸업은 해야지 라는 혼란스러운 마음이었다. 졸업은 마무리까지 고된 일임을 일 년간 여실히 느꼈다. 많은 졸업생들이 그렇겠지만 몸과 마음의 힘듦을 일 년 동안 가지면서 내년에 졸업할까 라는 생각이 매주, 매일 들었던 한 해였다. 주변인들의 착실한 도움 덕에 몸과 마음을 지켜내며 마무리 할 수 있었다. 그 많은 고마움들을 잊지않고 있음을 알려주고 싶다.

올해의 작업은 뜬구름처럼 다가왔던 작업과는 조금 달랐으면 했다. 그럴싸한 이야기들과 어려운 단어들, 있어 보이는 것에 회의감이 들었던 때였던 것 같다. 먼 미래나 먼 과거의 이야기가 아닌 현재와 나에게 집중한 감정들이 묻어났으면 했다. 더 많은 사람들이 공감하고 나조차도 이런 공간이 있으면 좋겠다 라는 진심이 생길 수 있는 프로젝트를 하고 싶었다. 목표와 가까워지고 있는가 반문하면서 일 년을 보내니 때로는 확신이 들면서도 가끔은 후회와 의심이 들어오는 순간이 있었다. 전시를 끝내고 방문록을 읽으면서 어느 정도 대답을 들은 것 같다. '그 찬란한, 삶의 영원할 기억이 실제 이런 공간에서 일어나면 좋겠습니다.' 라는 글을 남긴 익명의 사람에게 위로를 얻고 마무리할 수 있어 감사하다.

Magnetic Field City

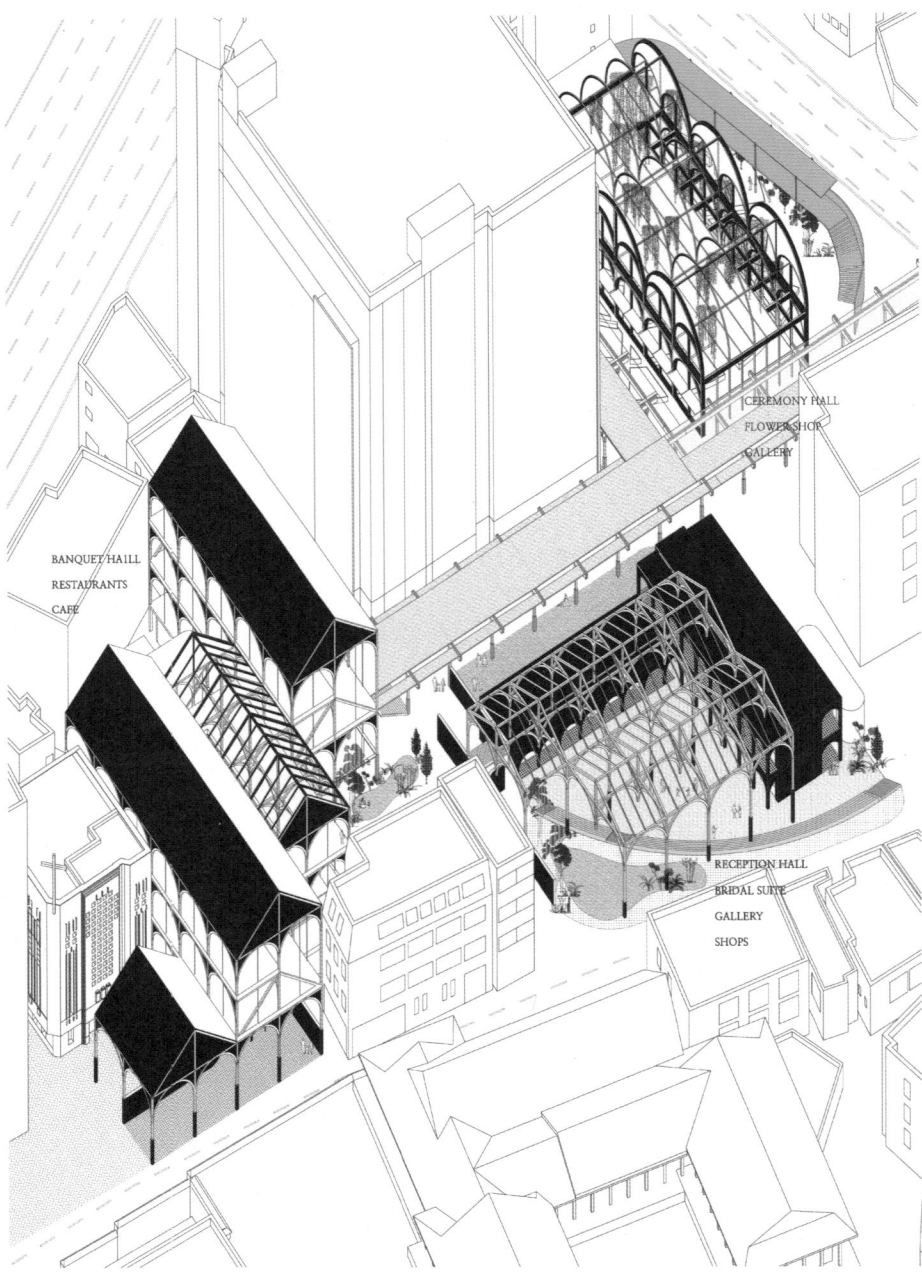

The direction of the studio was to find out how users enjoy cities and architecture from their perspective and to start an urban research by borrowing 'play' for new architectural plans. Sudoku was adopted to resolve the rigidity of the city and to find an urban plan with a new flow. I noted that in Sudoku, the process of searching for numbers to fill in the gaps was a repetition of temporal flows that crossed the boundaries of the axis of the Sudoku board. We replaced the temporal flows of Sudoku with moments of ephemeral connections within the city. Hearing someone cheer in a bar and cheering along with them is an ecstatic blurring of boundaries, experienced through an ephemeral bond with a stranger, a stranger occupying an entirely different location. This is the ephemeral flow, the fleeting yet lingering bonds that we find in cities.

As a program for urban space that could express this ephemeral bond, I explored the way weddings can penetrate the city. Weddings are a collection of acts where people gather, celebrate, and eat together in a space we call a wedding hall. By breaking up these sequential processes, each act takes place at a 'magnet'; each maintaining temporal relationships from different locations. A 'magnet' is defined as a space that concentrates people, a space such as a stage. The magnets are a collective assembled by segmented parts of the program, retaining an independent program while simultaneously maintaining a reciprocative relationship with neighbouring magnets.

In order to maintain the magnetic field of the magnets, which in turn blurs the boundaries such as day and night, weekday and weekend, Pil-dong creates the possibility for various program variations to occur within one magnet. Structurally, this is achieved through arch-shaped modules which develop a frame structure, which in turn forms large spaces to accommodate such possibilities. Pil-dong's scattered magnets bestow moments of ephemeral bonds, and creates a city with healthy emotional consumption.

사용자의 관점에서 도시, 건축을 즐기는 방식을 찾고 새로운 건축계획을 위해 '놀이'를 차용하여 도시 리서치를 시작하는 것이 스튜디오의 방향이었다. 도시의 경직성을 해소하고 새로운 흐름을 가진 도시계획을 찾기 위해 스도쿠를 채택하였다. 빈칸의 숫자를 찾아나가는 과정은 스도쿠판 축의 경계를 가로지르는 일시적인 흐름의 반복이라는 점에 주목하였다. 스도쿠의 일시적인 흐름은 도시에서 일시적인 유대감을 느끼는 순간으로 치환하였다. 주점에서 누군가의 환호성을 듣고 함께 환호하는 것은 일면식이 없고 다른 위치값을 가진 누군가와 순간의 유대감으로 경계가 흐려지는 즐거움을 느끼는 것이다. 이것이 도시에서 찾은 곧 사라지지만 여운이 남는 일시적인 흐름 즉, 일시적 유대감이다.

이러한 일시적 유대감을 발현할 도시공간의 프로그램으로 결혼식이 도시에 관입할 방식을 모색하였다. 결혼식은 사람들이 모여 축하하고, 함께 식사하는 행위가 집합되어 결혼식장이라는 공간에서 이루어진다. 이러한 순차적인 과정들을 분절시켜 각 행위는 다른 위치값에서 시간적 관계성을 가진 '마그넷'에서 일어나게 된다. 마치 무대공간처럼 사람들을 집중시키는 공간을 '마그넷'이라 정의하였다. '마그넷'은 프로그램의 과정들을 분절시켜 재조립한 집합이며, 독립적인 프로그램을 가지면서 다른 마그넷과 상호작용적인 관계를 가진다.

필동의 낮과 밤, 주중과 주말이라는 경계를 흐리는 마그넷들의 자기장을 유지하기 위하여 하나의 마그넷 안에서 다양한 프로그램의 변주가 생길 수 있는 가능성을 만들고, 이를 수용할 수 있는 아치형태의 가구식구조의 모듈을 형성하여 대공간을 구성한다. 필동의 흩뿌려진 마그넷은 일시적 유대의 순간을 선사하고 건강한 감정 소모를 하는 도시를 만든다.

1. SUDOKU WITH CITY
    – Sudoku ephemeral movement
    – Ephemeral intimacy
    – The rigidity of existing urban planning

2. RESEARCH
    – Urban trails
    – Stages

3. PROPOSAL
    – Programs
    – Magnetic fields

4. ARCHITECTURAL DRAWINGS
    – Plans
    – Axone diagram
    – Perspective drawings

Albeit I've never experienced the atmosphere and sentiments of old villages of the past portrayed in dramas and movies first hand, I understood the feeling - which always left me slightly envious and disappointed. I could imagine the townspeople gathering to eat and drink, and passer-by's peeping over the fence and joining in. It's a vague envy of the warmth of a time I've never experienced, and I felt the need to bring something from the past into a city that always seems to be missing something. Although times have changed, I felt that those moments of celebrating someones good news and being together as one was still relevant. Thus this project started upon a memory of a village feast in the past.

I started by exploring the possibilities of new urban planning through Sudoku, and then conducted research to find elements that creates and blurs the boundaries of cities while Sudoku-izing them. A solution to blurring the cities boundaries was suggested through program recombination techniques. The boundaries of the city occurring in Pil-dong were defined as the big differences in population depending on the day and the time. This difference was minimized by utilizing the different occupancy times of weddings and flower markets. Each separate space in the program was defined as a "magnet" due to its inherent power of concentrating people and connecting experiences as these people move through each space.

I noted that the dense, low-scale buildings behind the large buildings facing the boulevard were not easily accessible and had a depressed atmosphere due to the relatively tall and large buildings, so I selected these areas as targets to find out how and in what form the magnets would be placed. Based upon the program sequence and site research, the massing and placement scheme were developed. Concluding the project, to properly integrate the designed buildings within the site, select floors and the overall elevation of the Daehan Theater was renovated.

드라마와 영화에서 보여준 과거 옛 마을의 분위기와 정서는 겪지 않았지만 이해가 되었고, 그렇기에 부러움과 아쉬움이 있었다. 경조사에는 마을 사람들이 모여 음식을 하고 동네를 지나가는 사람들도 담장 너머 구경하며 함께하는 모습을 상상할 수 있다. 경험하지 못한 시대의 따뜻함에 대한 막연한 부러움이며, 아쉬움을 주는 지금의 도시에 과거의 무언가를 넣을 필요가 있다고 느꼈다. 시대는 변했지만 누군가를 축하하며 공통된 마음을 가지는 그 순간들은 아직 유효하다고 보았고, 이 프로젝트는 과거로 회귀하여 옛 마을 잔치가 벌어지는 모습을 상상하며 시작하였다.

스도쿠를 통해 새로운 도시계획의 가능성을 찾아보는 작업을 시작으로 하여 도시를 스도쿠화하면서 도시의 경계를 만드는 요소와 흐리는 요소를 찾는 리서치를 진행하였다. 프로그램의 재조합 방식을 찾아 도시의 경계를 흐리는 해법으로 제시하였다. 필동에서 나타나는 도시의 경계는 요일별, 시간별 생활인구의 큰 차이라고 보았고, 결혼식과 꽃시장의 다른 점유시간대를 이용하여 차이를 줄이고자 하였다. 프로그램이 분리된 각각의 공간들은 사람들을 집중하는 힘이 있으며 각각의 공간을 이동하며 경험의 연결이 발생할 수 있으므로 '마그넷'으로 정의하였다.

대로변에 면하는 큰 건물들 배면으로 조밀하고 낮게 구성된 건물들은 비교적 높고 규모가 큰 건물들로 인해 접근이 쉽지 않고 침체된 분위기를 갖는다는 점에 주목하였고, 이러한 곳을 대상으로 선정하여 마그넷이 배치될 방식과 형태를 찾아나갔다. 프로그램의 순서와 사이트 리서치를 기반으로 매스 디벨롭과 배치를 고려하였고, 설계한 건물들과 함께 대상지가 잘 작동하기 위해 대한극장의 일부 층과 입면을 리노베이션하면서 프로젝트를 마무리하였다.

Magnetic Field City

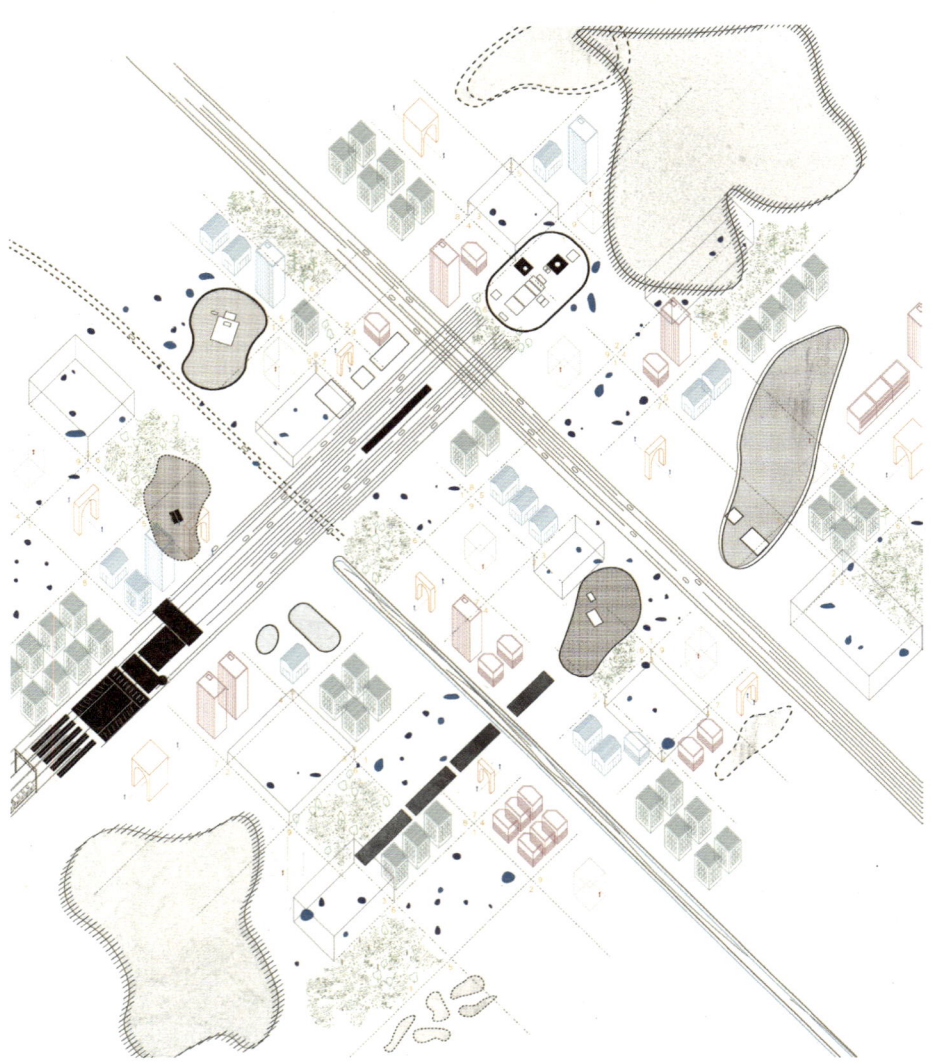

The point of interest in Sudoku is the flow across axes that occurs as you fill in the blanks. The flow of blue dots on the Sudoku board is the flow that occurs to find a number in a space. This is a "temporary flow" that disappears when another number is found, and these flows are repeated to complete the puzzle. The phenomenon of temporary connections between two numbers in different positions in a Sudoku board is also found in cities. When you hear someone cheering in a bar, even if you don't know them, the moment of being together at different points of space is similar to the phenomenon of ephemeral bonding in the city. These ephemeral bonds add a little vigour to everyday life, and I believe we need urban spaces where such moments could occur.

스도쿠에서 주목한 지점은 빈칸을 채워나가는 과정에서 나타나는 축을 넘나드는 흐름이다. 스도쿠 판에 표현된 파란 점들의 흐름은 한 칸의 숫자를 찾기 위해 생기는 흐름이다. 이 과정은 또 다른 숫자를 찾을 때는 사라질 '일시적인 흐름'이고 이러한 흐름들이 반복되며 완성한다. 스도쿠에서 위치값이 다른 두 숫자가 일시적으로 연결이 생기고 사라지는 현상은 도시에서도 찾아볼 수 있다. 주점에서 누군가 축하하며 환호하는 소리를 들으면 일면식이 없더라도 함께하는 순간이 도시에서 일시적인 유대감이 생기는 현상과 유사하다고 보았다. 일시적인 유대감은 일상에 작은 활기를 주며 그러한 순간들이 발생할 수 있는 도시공간이 필요하다고 생각하였다.

I conducted research by Sudoku-izing the old city center around Gyeongbokgung Palace. From Inwangsan Mountain to Naksan Mountain, and from Bukaksan Mountain to Namsan Mountain, I looked for elements of Sudoku's axes and given numbers. I tried to represent the elements of the two axes that divide the area of the city as more permanent elements of the city, and the nine squares within them as zoning districts for buildings. Compared to the existing city, I thought that a certain flow in the Sudoku city could connect the city by crossing the two axes and the permanent elements.

I thought that the flows in the Sudoku city were similar to the flows of the hiking routes that naturally occurred in cities as they crossed the boundaries between mountains and cities. In order to find the elements that create the same flow in the city, I conducted research from the foot of Namsan Mountain to the vicinity of Gyeongbokgung Palace, marking the points that I recorded with photographs. I found elements that attracted people like magnets, such as buildings with a sense of time, surroundings and unusual elevations, nature in the city, monumental architecture, and the act of eventfulness on the street.

경복궁 일대의 구도심을 스도쿠화 하면서 리서치를 진행하였다. 인왕산에서 낙산, 북악산부터 남산까지의 범위에서 스도쿠의 축들과 주어진 숫자들의 요소를 찾아보았다. 변화의 가능성이 적고 도시의 영역을 나누는 두 축의 요소와 그 안의 나누어진 9칸을 건물의 용도 지역지구로 표현해보았다. 기존 도시와 비교했을 때, 스도쿠 도시에서의 보여지는 어떠한 흐름이 두 축과 주어진 요소들을 넘나들며 도시를 엮어줄 수 있다고 생각했다.

스도쿠 도시에서 보여지는 흐름은 도시에서 산과 도시의 경계를 넘나들며 자연스럽게 생겨난 둘레길의 흐름과 비슷하다고 생각했다. 도시에서 둘레길과 같은 흐름을 만드는 요소를 찾기 위해 남산자락에서 경복궁 부근까지의 큰 목적지만 두고 사진으로 기록한 지점들을 표시하면서 리서치를 진행하였다. 시간성을 담은 건물, 주변과 이색적인 입면, 도심 속의 자연, 기념비적인 건축물, 거리에서 일어나는 이벤트성의 행위 등 자석처럼 발걸음을 향하게 하는 요소들을 찾아볼 수 있었다.

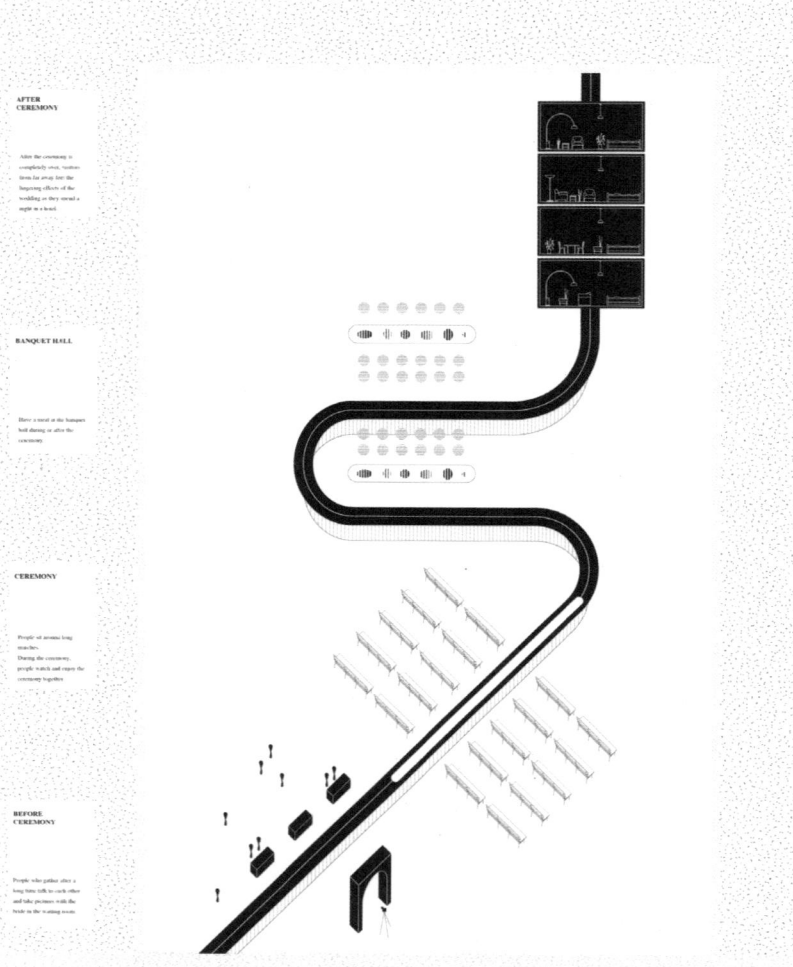

Pil-dong, Jung-gu, was selected as the target area, and through census data collected at four locations, we found that in Pil-dong, where there are many offices, weekdays are much more active than weekends, and there are more people during the day than at night, showing the phenomenon of urban communalization. If a magnet that attracts people's attention is placed on the site, which is an element that creates flow by Sudokuizing Pil-dong, as shown in the urban Sudoku, the surrounding area can create a connection that crosses boundaries, akin to magnetism.

대상지로 선정한 중구 필동은 4개의 지점에서 조사된 생활인구 데이터를 통해 오피스가 많은 필동의 평일이 주말보다 활발하고 낮시간이 밤보다 더 많은 인구가 현주해 있는 도시공동화 현상이 보여지는 곳임을 확인했다. 도시 스도쿠화로 나타낸 것과 마찬가지로 필동을 스도쿠화하여 흐름을 만드는 요소인, 사람들의 관심을 끌어내는 마그넷이 사이트에 배치된다면 그 주변부는 자력처럼 경계를 넘나드는 연결을 만들어줄 수 있다.

Magnetic Field City

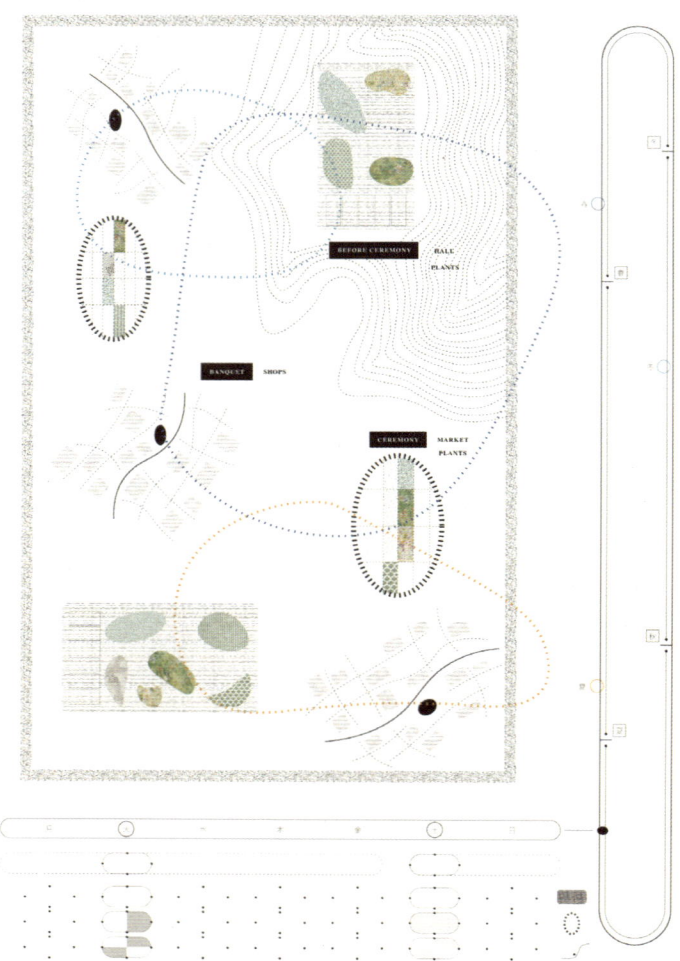

Magnets, urban spaces that gather people, were conceived as connected programs that could have their own independent existence while simultaneously maintaining an interactive relationship with one another. A wedding, which brings together a large number of people, was seen as an assemblage of temporal relationships with various programs in the city. If you break down this sequence of people coming together in one building, watching the ceremony, celebrating and eating together, and then leaving, you have programs that are separate in space but have a temporal sequence, so they can be independant while staying connected; thus working like magnets. And if a wedding, an event where usually only invitees can experience, could be experienced by people wandering through the city, Pildong could be energized by moments of ephemeral bonds.

In addition to active weddings on weekends, the flower market was shown to be a program that would serve as a magnet when there is no ceremony, as it is filled from morning to night, has separate spaces for different functions, and is highly utilized when accompanying weddings. I thought that if there were such large spaces scattered around the city where many people gathered, it could be a magnet for not only weddings, but also events to gather, watch, and celebrate.

사람들을 모으는 도시공간인 마그넷들은 각자 독립적으로 존재하는 프로그램을 가지면서 상호작용적인 관계를 유지할 수 있는 연결성이 있는 프로그램을 생각했다. 다수의 사람들이 모이는 결혼식은 도시의 다양한 프로그램을 지니는 동시에 시간적 관계성을 가진 집합체로 보여졌다. 한 건물 안에서 사람들이 모여 식을 보고, 함께 축하하며 식사를 한 후 돌아가는 이러한 일련의 과정들을 분절시킨다면, 공간은 분리되어 있지만 시간적 순서를 갖는 프로그램이므로 마그넷처럼 연결되며 작동할 수 있다. 또한 특정 공간에서 초대된 사람들만 경험할 수 있는 결혼식을 도시를 거닐면서 마주할 수 있다면 일시적인 유대감을 느끼는 순간들로 필동에 활기를 불러일으킬 수 있을 것이다.

주말에 활발한 결혼식과 더불어 식이 없을 때 마그넷 역할을 할 프로그램으로 꽃시장은 아침부터 밤의 시간이 일정하게 채워져있고, 각 기능이 다른 공간들이 분리되어 있으며 결혼식과 함께 있을 때의 활용성이 높은 프로그램으로 보여졌다. 이처럼 많은 사람들이 모이는 대공간이 도시에 흩뿌려져있다면 결혼식뿐만 아니라 모이고 구경하고 기념하는 이벤트를 담아내는 마그넷이 될 수 있을 것이라 생각했다.

Magnetic Field City

One place where the high and low foot traffic in Pil-dong was clearly visible was in the parking lots. The parking lots are located in areas surrounded by offices and restaurants, and there was at least one parking lot within 100 meters of each other. On weekends, when the parking lots are less crowded, they become empty, and we thought that placing magnets in these locations would change the aesthetics of the city.

I placed the magnets by predicting weekend and weekday traffic, with and without weddings. Weddings, markets, and other events all involve a similar process of people gathering, watching, celebrating, and eating. Rather than circling around a single building, the magnets scattered around the city create an experience of moving through the city, celebrating someone and having the joy of being together celebrating a common cause. Different types of stages with independent programs can be scattered around the city, increasing the frequency of ephemeral bonds.

필동에 유동 인구가 많고 적음이 뚜렷하게 보여지는 곳은 주차장이었다. 주차장은 오피스와 식당으로 둘러싸인 곳에 위치하며, 지상에 100미터 내 하나 이상의 주차장이 존재하고 있었다. 사람이 적은 주말의 주차장은 비어있는 공간이 되며, 이러한 위치에 마그넷을 넣으면서 도시 미관도 변화할 수 있다고 생각했다.

결혼식이 있을 때와 없을 때 주말과 주중의 동선을 예측하여 마그넷을 배치하였다. 결혼식, 시장, 여러 행사들 모두 사람들이 모이고 구경하고 기념하고 식사하는 일련의 과정은 어디서나 유사하다. 이런 과정을 하나의 건물에서 쳇바퀴처럼 돌아가는 것이 아닌, 도시에 분산된 마그넷으로 도시를 유영하며 누군가를 축하하고 함께 공통된 마음을 갖는 즐거움을 얻는 경험을 선사한다. 독립적인 프로그램을 가진 다양한 형태의 무대들이 도시에 흩뿌려지면서 일시적인 유대감의 빈도를 높일 수 있다.

Mass Study — Bogyong Kim

Pil-dong is divided into three administrative districts; a general commercial area, the third class general residential area, and the second class general residential area. Behind the tall and large buildings facing the main road, low buildings under 7 stories are densely arranged. The backside of the Daehan Theater populated by empty buildings and stagnant spaces, ultimately became the target site. It is a stagnant area with limited visual information due to the many parking lots and large buildings in Pil-dong. While changing the image of the back, we considered a structure that can accommodate temporary and variable programs. A single arch module is utilized, adjusting its scale accordingly. The structure can be diversified with braces and walls, giving a sense of openness to the large space.

필동은 일반상업지역, 제3종일반주거지역, 제2종일반주거지역으로 나누어져 있다. 큰 도로와 면하는 높고 큰 건물들 뒤쪽으로 7층 이하의 건물들이 밀도있게 배치되어 있는데, 빈건물들이 많고 침체되어있는 곳이 대상지로 정한 대한극장의 배면이다. 필동의 많은 주차장 건물과 큰 건물들로 인해 시각적인 정보가 제한되면서 침체되어 있는 구역이다. 배면의 이미지를 바꿔주면서, 일시적이면서 가변적인 프로그램을 수용할 수 있는 구조를 고민하였다. 하나의 아치모듈이 규모에 따라 변화가능하고 가세와 벽으로 다양성을 줄 수 있는 구조이며 대공간의 개방감을 줄 수 있도록 하였다.

Following the Korean house's accessible stone walkway, the walkway leads to a large hall with a lowered ground level. During the ceremony, the reception desk is on the left, leading to the bridal suite at the back of the hall, with a garden in the background.

A long corridor leads to the greenhouse where the ceremony takes place. The interior botanical garden is divided by chairs and pathways, and connected naturally to the outside. The chairs are arranged in an arch that spreads around the dais. In addition to the plant shop on the first floor, the second floor is used as a gallery and rest area. A bridge on the right side allows guests to view the ceremony and exhibits related to the shop.

The banquet hall along the corridor includes the loading and unloading area, and the kitchen. Service areas such as the loading and unloading area, kitchen, and staff rest room are located at the back. When the restaurant is not used for weddings, the entrance is arranged in consideration of the main access route along the stone wall. After entering the main entrance, there is a circular staircase and garden in the center, and the circular staircase leads to the third floor. The elevator, which is accessible from the outside, is arranged in a vertical line for going up and down.

Small gardens have been created between the existing buildings and the new buildings, and you can enjoy the view from the windows and stay in the garden. In addition, the first floor of the Daehan Theater, located between the main street and the inner alleys, was opened up for easy access and renovated to resemble visually and structurally, the surrounding arched elevations.

The corridor is a passageway that connects the Daehan Theater and the three buildings on two axes, and wanting the structure to be different from the surrounding existing formal language. The roof was divided into three plates differing in level and material, to create a sense of openness between the buildings.

사람들의 접근성이 높은 한국의 집 돌담길을 따라 들어오면 지면의 레벨이 낮아지는 대공간 홀로 접근할 수 있다. 식이 있을 때는 좌측에 위치한 리셉션 데스크를 지나 홀 안쪽에 위치한 신부대기실로 연결이 되고 뒤쪽으로는 신부대기실의 배경이 되어주는 정원이 있다.

긴 회랑을 따라 이동하면 식이 열리는 온실이 위치해있다. 내부의 식물정원은 의자와 통로로 구획이 생기고 외부와 자연스럽게 연결되었다. 의자들은 단상을 중심으로 퍼져나가는 호의 형태로 배치되었다. 1층의 식물상점과 더불어 2층은 갤러리, 휴게공간으로 사용된다. 우측에 접해있는 브리지에서 식을 구경하거나 상점과 관련된 전시도 가능하다.

회랑을 따라 이어진 연회장은 상하차구역, 주방, 직원휴게실 등 서비스 동선을 뒤쪽으로 배치하였다. 결혼식이 없는 평소에 식당을 이용할 때는 돌담길 방향의 주 진입 동선을 고려하여 입구를 배치하였다. 주출입구로 들어서면 중정의 원형계단과 정원이 있고 원형계단은 3층까지 연결된다. 상하차를 위한 수직동선으로 외부에서 접근가능한 엘리베이터가 배치되어 있다.

기존 건물들과 새로 들어선 건물들 사이에는 작은 정원들이 조성되어 있으며 창밖으로 조망하고 정원에 머무를 수 있다. 또한 큰 길과 안쪽 골목들 사이에 위치한 대한극장의 1층을 열어주면서 진입이 용이하도록 하였고 주변의 아치형태의 입면들과 유사한 입면을 가지도록 리노베이션을 진행하였다.

회랑은 두 축으로 연결되어 대한극장과 세 건물 사이를 연결하는 통로이며 주변의 형태적 언어와 다른 구조였으면 하였고, 지붕은 세 개의 판으로 나누어 레벨과 재료의 차이를 주어 건물들 사이 공간의 개방감을 주려고 하였다.

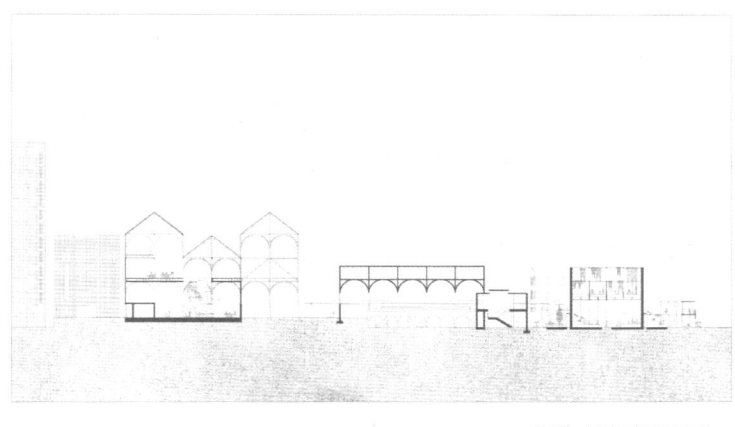

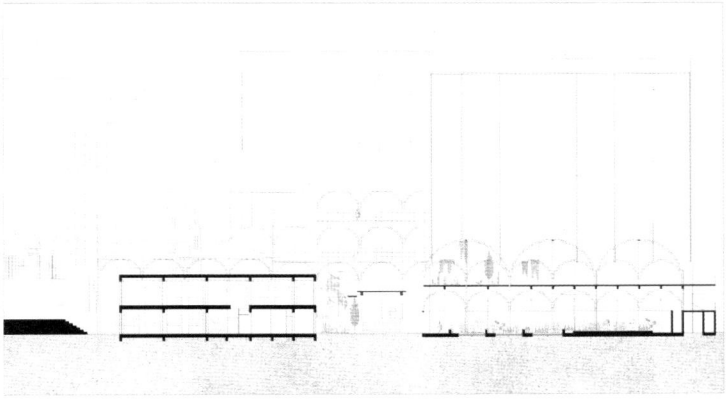

Magnetic Field City

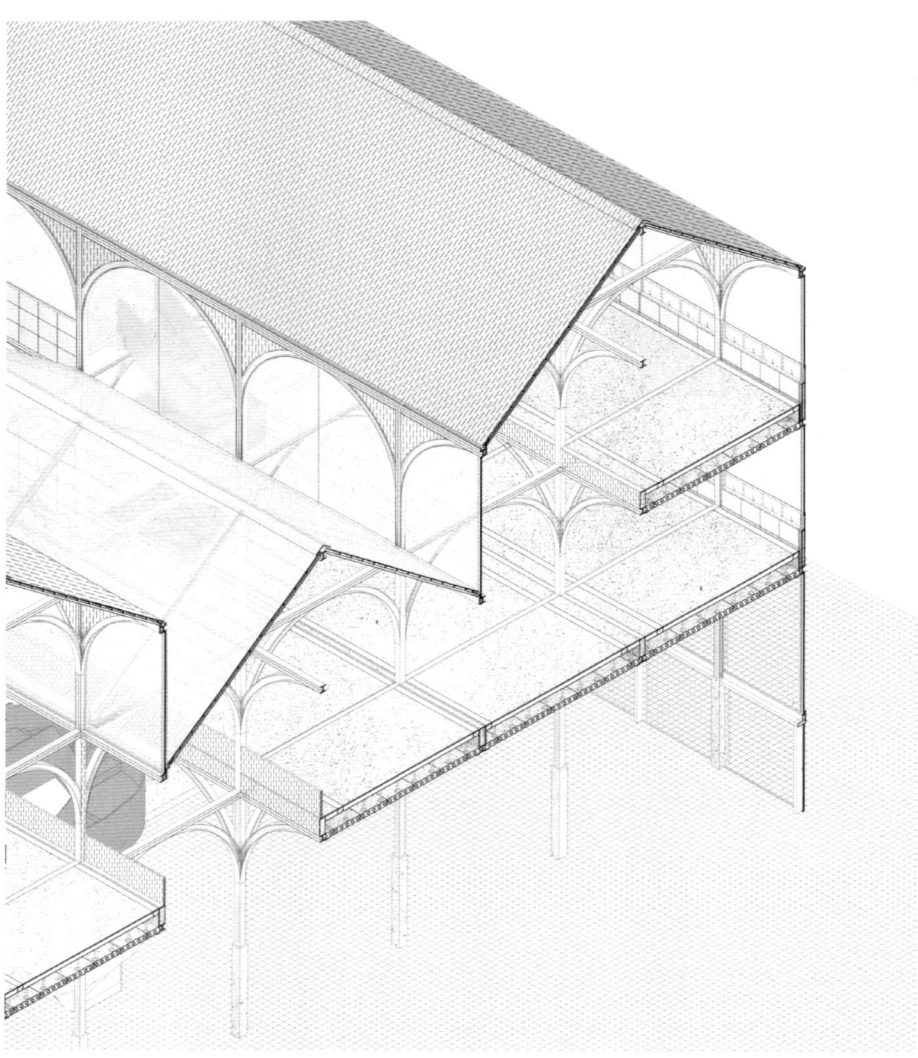

Due to the nature of the program, I felt that the arch structure would set the tone for the space. Rather than a heavy arch, we decided that a light and open arch would be an appropriate solution in contrast to the dense site, and implemented the arch form as a module for a frame structure.

I created a 6-meter structural module with a steel column structure, with trusses and arches as bracing. Since the program was not fixed, but variable and open to change at any time, the structure was also made of modules that could facilitate such change. Some of the modules on the elevations have fixed windows in the curves of the arched openings, with ventilation windows underneath, in order to avoid disrupting the structure of the arches from the inside.

The three gabled roofs allow daylighting by translucentizing the material of the roof at the center, and air circulation by opening the area where the two sides of the roof meet.

프로그램의 특성상 아치 구조가 공간의 분위기를 만들어줄 것이라 생각했다. 무게감이 느껴지는 아치보다는 건물의 밀도가 높은 사이트에서 가벼운 느낌의 개방감이 있는 아치형태가 적합하다고 보았고 아치형태를 가구식구조로 풀었다.

철골 기둥보구조에 트러스와 아치가 브레이싱 역할을 하는 6미터 구조 모듈을 만들었다. 고정된 프로그램이 아닌 언제나 변화할 수 있는 가변적인 프로그램이므로 구조 또한 모듈로 이루어져 변화가 용이할 수 있도록 하였다. 내부에서 아치형태의 구조를 해치지 않기 위해 악세스플로어로 설비를 넣은 방식을 채택하였다. 또한 입면의 일부 모듈은 아치형태의 개구부 곡선 부분을 고정창으로 두고 그 아래쪽에 환기창을 두었다.

박공형태의 세 개의 지붕은 중정이 위치한 지붕의 재료를 반투명하게 하여 일조량을 확보할 수 있도록 하였고, 지붕의 두 면이 만나는 부분을 열어줌으로써 공기의 순환이 가능하게 하였다.

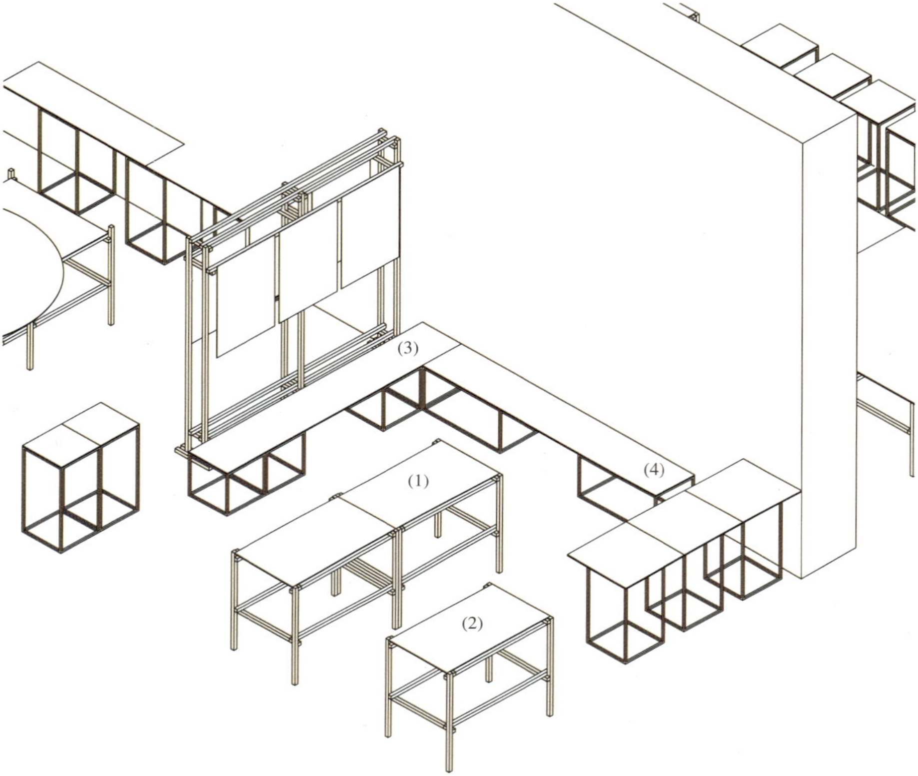

Magnetic Field City

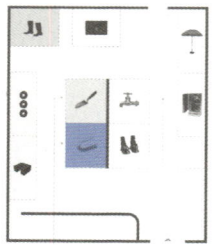

1. Main Model *4
2. Site Model
3. Structure Model
4. Concept Model

졸업도우미　현은우
　　　　　　김지후. 김동현. 신재형.
　　　　　　최카밀라. 최시원.

MAIN MODEL

Scale 1:100
Laser cut royal board, 100 by 140 cm

Magnetic Field City

SITE MODEL

Scale 1:1500
Woodrock & 3D print, 40 by 75 cm

STRUCTURE MODEL

Scale 1:100
Laser cut royal board, 50 by 50 cm

Rain Festival　　　　Taeyoung Kim Studio

김서진
Seojin Kim

Rain Festival

1-1 전시사진

Preface

Seojin Kim

Rain Festival is a story that began as a reflection on living with plants. The experience of first bringing plants into a small room was expected to be extended to the city. The inequalities caused by the climate crisis soon turned out to be social inequalities, and the hope was that neighborhoods with vulnerabilities to rain could overcome them and become communities that cared for each other.

There are certain things that need to be done every season, every year, and I believed in the transformation that happens when they become routine. It is as if they spent a year with a small belief that they would feel a tangible change in their repetitive and endless routine.

Along with the empathic drawings that originated from personal experience, the exhibition includes climate reports, including figures that objectively confirm recent climate anomalies, and preliminary research materials prepared for the process of building an addition to an existing building. In addition, it was a personal pleasure to observe and record the characteristics of the current residents and their lives season by season, along with the spatial analysis of the existing building. In addition, a short video was made to accompany the exhibition as a way to show the lives of the people and their plants in Beon-dong by combining articles from the beginning of the special type of permanent rental apartment in Beon-dong with observations of the people and their plants for a year.

It's a year that I wouldn't dare say is short, but if you think of it as a project, it still leaves something to be desired. However, I would like to value the effort I put into refining the story to convey my favorite things over the course of the year. It's raining a lot this year. I hope that many people have a similar feeling on a rainy day and finish the graduation exhibition with the expectation of becoming a neighbor who reaches outward like a plant.

'Rain Festival'은 식물과 함께 하는 삶에 대한 고찰에서 시작된 이야기이다. 작은 방에 식물을 처음 들였을 때의 경험이 도시로 확대되는 것을 기대했다. 기후 위기로 인한 불균등은 곧 사회적 불평등으로 드러났고, 비에 대한 취약함을 가진 지역이 그것을 극복하고 서로를 돌보는 공동체가 되는 바람을 담았다.

매년 계절별로 해야 하는 일들이 있고 그것이 일상이 되었을 때의 변화를 믿었다. 반복적이고 끝없는 일상에서 그들이 감각적인 변화를 느끼게 될 것이라는 작은 믿음으로 1년을 보낸 듯 하다.

전시는 개인적인 경험에서 시작한 공감각적 드로잉과 함께, 최근 이상 기후를 객관적으로 확인할 수 있는 수치들을 포함한 기후 리포트와 기존의 건축물에 증축하는 과정을 위해 준비했던 사전 조사 자료들을 볼 수 있다. 또한 기존 건물의 공간적 분석과 함께 현재 살고 있는 사람들의 특성 그리고 그들의 삶을 계절별로 관찰하고 기록하는 일은 개인적으로도 즐거운 일이었다. 그에 더해 번동의 영구 임대 아파트라는 특수한 유형이 시작될 당시의 기사들과 1년 동안 번동의 사람들, 그리고 그들의 식물들을 관찰한 것을 엮어 그들의 삶을 보여줄 수 있는 방식으로서 간단한 영상도 함께 제작하였고, 전시와 함께 볼 수 있도록 준비했다.

감히 짧다고 말할 수 없는 1년이지만 프로젝트만 생각한다면 여전히 아쉬운 부분도 남아 있다. 그러나 1년 동안 좋아하는 것들을 주제로 전달하고 싶어 이야기를 다듬었던 노력들을 남긴 것에 가치를 두고 싶다.

올해도 비가 많이 내린다. 많은 사람들이 비가 오는 날 비슷한 마음을 가졌으면 좋겠다는 마음으로, 식물처럼 바깥으로 손을 뻗는 이웃이 되기를 기대하며 졸업 전시를 마친다.

Main Drawing

Seojin Kim

In a world where inequality is visible and discrimination abounds, society is constantly drawing lines and ranking each other. At a time when the climate crisis is acceler-ating yet another imbalance, I hope to transform the relationship between rain and architecture to restore individual well-being and community. A rainy day is no longer a catastrophe. Instead of fear, we are excited by it. Rainy days will become a festival that brings us together.

Residents of the first neighborhoods where permanent rental apartments were introduced are unable to leave the neighborhood for economic and physical reasons. Those who are trapped by the word 'forever' have stayed in Bun-dong since the 90s until now, and for them, it is life with plants that can make them feel the passing of time and wait for the repeated seasons.

Their innate terrain, starting from the valleys of the two mountains, made the rainfall faster and they were vulnerable to it. I want to ask for architectural changes to rainwater so that they are not afraid of rain, and their daily life is permeated by plants, so that they can wait for tomorrow when flowers bloom.

Architectural elements are transformed to respond to rainwater, filling the space between us with plants and water. By bringing what the plants need into the house, our senses are reawakened in the plant cycle, and we can become one with the plant community. Rainwater becomes the medium that brings the city together, and we now wait for the rain together.

Connected by water and plants, we become a new community that takes care of each other. The single constructions may be small, but they are solidarity, they connect the city as a system, and their power is powerful. The plant communities born from the restoration of equitable rainwater are sensory and alive. Now we, the people, wait for the rain.

불평등은 가시화되고 차별이 넘쳐나는 세상이다. 끊임없이 선을 긋고 서로의 순위를 매긴다. 기후 위기로 또 다른 불균형이 가속화되는 이 시점에 비와 건축의 관계의 변화를 도모하여 개인의 행복과 공동체의 회복을 기대한다. 비가 오는 날은 더 이상 재앙이 아니다. 우리는 두려움 대신, 함께 설렘을 느낀다. 비가 오는 날은 우리를 하나로 만드는 축제가 될 것이다.

영구 임대 아파트가 최초로 도입된 번동의 주민들은 경제적, 물리적 이유로 번동을 벗어날 수 없다. '영원'이라는 단어에 갇힌 사람들은 90년대부터 지금까지 번동에 머물렀고, 그들에게 흘러가는 시간을 느끼게 하고 반복되는 계절을 기다리게 할 수 있는 것은 식물과 함께하는 삶이다.

두 산의 골에서부터 시작되는 타고난 그 지형은 빗물을 더욱 빠르게 만들고, 그러한 빗물에 대응하기에 그들은 취약했다. 비를 두려워하지 않도록 빗물에 대한 건축적 변화를 요구하고, 그들의 일상을 식물들 속에 스며들게 하여 꽃이 피는 내일을 기다리게 하고 싶다.

건축적 요소들은 빗물에 대응하기 위해 변형되고, 우리 사이를 식물과 물로 채워나간다. 식물에게 필요한 것들을 집 안으로 들이며 식물의 사이클 속에서 우리의 감각은 다시 일어나고, 우리는 식물의 군집처럼 하나가 될 수 있다. 빗물은 도시를 하나로 만들어주는 매개체가 되고, 우리는 이제 함께 비를 기다린다.

물과 식물로 연결된 우리는 서로를 돌보는 새로운 공동체가 된다. 단일의 건축은 미미할 수 있으나 그들은 연대하고 도시를 하나의 시스템으로 연결되고 그 힘은 강력하다. 평등한 빗물의 회복으로 태어난 식물 공동체는 감각적이고 살아 있다. 이제 우리, 비를 기다리자.

1. 도입
    - 식물 감각
    - 도시 공공재로서의 식물
    - 기후, 식물 그리고 인간
    - 기후 위기와 녹색 불평등

2. 전개
    - 시나리오
    - '느린 물(slow water)'을 통한 건축적 제안
    - 서울의 기후 분석
    - 빗물에 가장 취약한 지역 선정
      사이트: 강북구 번동
    - 사이트 분석
    - 건축 유형별 분석 1: 아파트 유형
    - 건축 유형별 분석 2: 다세대 주택

3. 제안
    - 빗물 시스템 분석 및 제안
    - 아파트 리노베이션 방식 구축
    - 다세대 주택으로의 도입
    - 기존의 도시 인프라와의 연결을 통한
      시스템 구축
    - 식물과 빗물을 통한 돌봄 공동체 제안

Process — Seojin Kim

Bringing plants into your home means more than you might think. Growing plants is like bringing sunlight, wind, and soil into your home. And in our personal lives, we take actions to be with them. We water them, we open the windows, and we have a different life than before. The habit of plants to reach outward in search of sunlight gives me hope that it will reach beyond the fence to the person next door.

Everyone in the city grows plants. Together, in the park or on the sidewalk, even the wild flowers that suddenly bloom, rainwater will be the most necessary and equitable water we have, and we will all wait for it together.

At a time when rainwater is losing its equality due to the climate crisis, architecture can recreate the 'we' by rebuilding the relationship with water that it once had. Discrimination and inequality, which were taken for granted by capitalism, are no more, and we all wait for the rain with one mind, and with that mind, through the act of growing plants, we become one.

Bun-dong, Gangbuk-gu is divided into two main architectural types. Multi-family houses on hills and permanent rental apartments on flat land. The topographical characteristics of the two types determine the speed of rainfall and require different conditions. Typologizing the existing architecture and seeking change from it is also the hope that they can become a plant community that respects their current life, but also takes care of each other enough. For those who are unable to leave the downtown for economic reasons, the hope is that by focusing on elements that can bring sensory change and anticipation for the future to their lives, they can live a life filled with anticipation rather than boredom of 'repetition.'

식물을 집으로 들인다는 것은 생각보다 많은 의미를 지닌다. 식물을 키우는 것은 집 안으로 햇빛과 바람, 흙을 들이는 것과 같다. 그리고 개인의 삶에서 그것들과 함께 하기 위한 행동들이 발현된다. 물을 주고, 창문을 열며 우리는 이전과는 다른 삶을 가지게 된다. 햇빛을 찾아 바깥으로 뻗어나가는 식물의 습성은 나에게 담장을 넘어 옆집의 사람에게도 닿지 않을까 하는 희망을 느끼게 했다.

도시의 모든 사람들이 식물을 키운다. 공원에서 혹은 도로에 갑자기 피어난 들꽃까지 함께 가꾼다. 그 때 빗물은 우리에게 가장 필요하고 공평한 물이 될 것이며 우리는 모두 함께 비를 기다릴 것이다.

기후 위기로 빗물이 평등함을 잃어가는 지금, 건축이 과거에 가졌던 물과의 관계를 다시 구축함으로써 '우리'라는 관계를 다시 만들어 갈 수 있다. 자본주의로 인해 당연히 여겨졌던 차별과 불평등은 이제 없으며 모두 한마음으로 비를 기다리며 그 마음으로, 식물을 키우는 행위로 우리는 하나가 된다.

강북구 번동은 크게 두가지의 건축 유형으로 나뉜다. 언덕에 위치한 다세대 주택과 평지에 위치한 영구 임대 아파트이다. 두 유형이 가진 지형적 특성은 내리는 빗물의 속도를 결정하며 각각 다른 조건을 필요로 했다. 기존에 있던 건축물을 유형화하고 그것으로부터 변화를 추구하는 것은 현재 그들의 삶을 존중하면서도 그들은 충분히 서로를 돌보는 식물 공동체가 될 수 있다는 희망이기도 하다. 경제적인 이유로 번동을 떠나지 못하는 사람들을 위해 그들의 삶에 감각적인 변화와 미래에 대한 기대를 불어넣을 수 있는 요소에 집중하여 '반복'에 대한 지루함이 아닌 기대감으로 가득 찬 삶을 희망한다.

Rain Festival

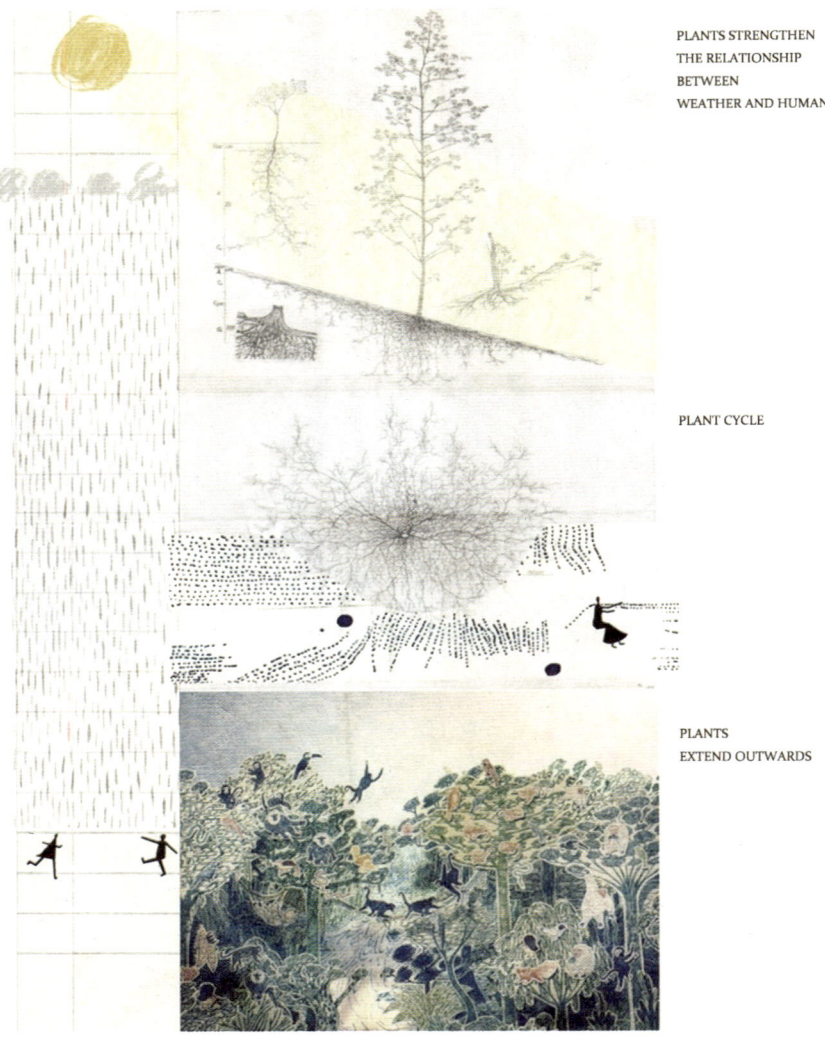

PLANTS STRENGTHEN
THE RELATIONSHIP
BETWEEN
WEATHER AND HUMANS

PLANT CYCLE

PLANTS
EXTEND OUTWARDS

1. Plants

Consider the sensory experience of growing plants. Our senses come alive with the different elements of growing a plant, and we begin to pay attention to the tiny leaves. Bring sunlight and breezes into the house. The most basic act of watering will become an act that allows us to structure our daily routine. Like plants reaching out for sunlight, we reach out to the outside world. The properties of plants have the power to transform our currently disconnected society.

2. Rainwater

Rain used to be our most equitable water source for maintaining urban greenery, but extreme weather events have led to inequalities in rainwater availability. GREEN INFRASTRUCTURE, once a public good, is now a measure of economic and social vulnerability. Just as California's water scarcity has led to a ban on growing lawns and people checking their own gardens, this will be taken to an extreme, with communities monitoring each other. Architectural changes to rainwater management and solidarity among people are needed to stop fearing rain as a disaster.

1. 식물

식물을 키울 때의 감각에 대해 고찰한다. 식물을 키우기 위한 여러 요소들로부터 우리의 감각이 살아나고, 작은 잎에도 신경 쓰기 시작한다. 집안에 햇빛과 바람을 들인다. 물을 주는 가장 기본적인 행위는 우리의 일상을 구조화할 수 있도록 하는 행위가 될 것이다. 햇빛을 찾아 밖으로 뻗어나가는 식물처럼 우리 또한 바깥으로 손을 뻗는다. 식물이 가지는 특성은 현재 단절된 우리의 사회를 변화시킬 수 있는 힘이 있다.

2. 빗물

비는 도시의 녹지를 유지할 수 있는, 어쩌면 우리에게 가장 공평한 물이었다. 그러나 이상 기후로 인해 빗물이 가져온 불균등은 곧 녹지의 불평등으로 이어졌다. 공공성을 가졌던 녹색 인프라(GREEN INFRA)는 이제 경제적, 사회적 취약성의 척도로 자리 잡게 되었다. 캘리포니아에서 물 부족으로 인해 잔디를 키우는 것을 금지하게 되면서 각자의 정원을 확인하듯, 이는 극단적으로 나아가 서로를 감시하게 되는 공동체로 번져갈 것이다. 재난으로서 비를 두려워하지 않기 위해서 빗물에 대한 건축적 변화와 그들 간의 연대가 필요하다.

Rain Festival

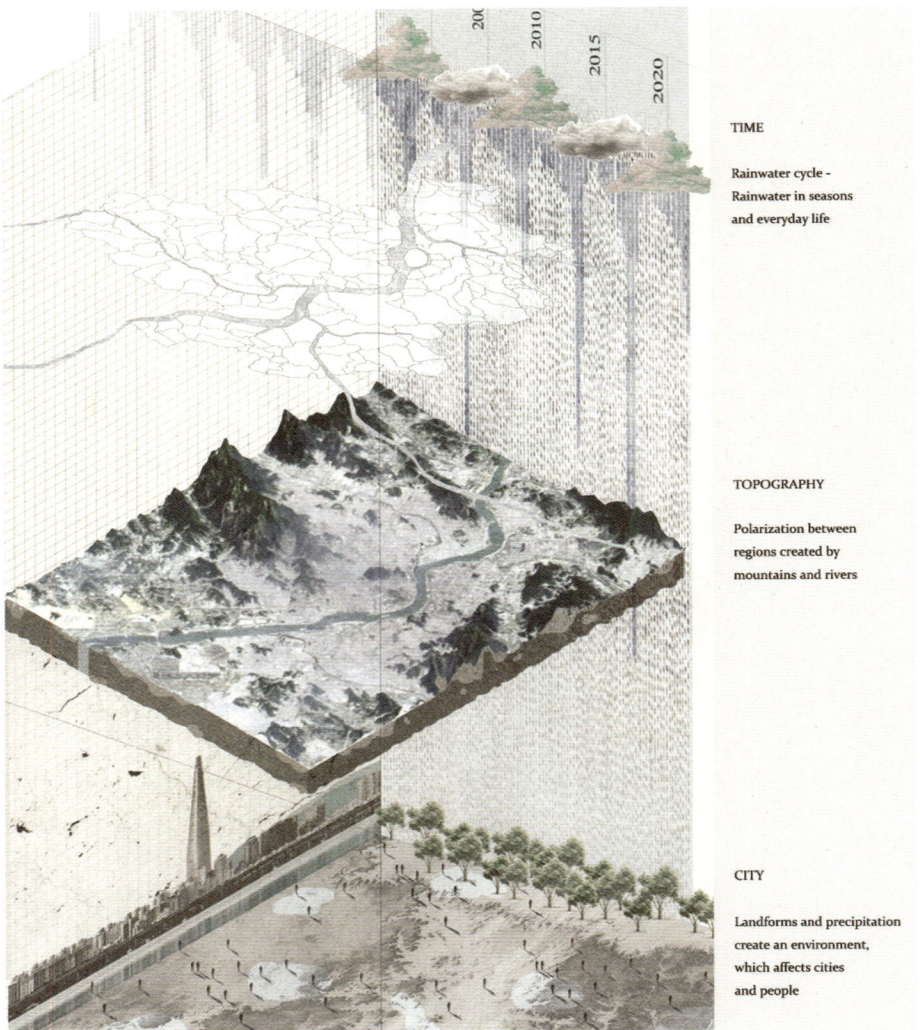

TIME

Rainwater cycle - Rainwater in seasons and everyday life

TOPOGRAPHY

Polarization between regions created by mountains and rivers

CITY

Landforms and precipitation create an environment, which affects cities and people

By examining 10 years of precipitation and vulnerability to it based on Seoul, I found that regional and seasonal polarization is intensifying due to extreme weather, and that the ability to respond to it is not only influenced by innate factors such as terrain and climate, but also by social indicators such as the proportion of elderly people and the proportion of basic water users, as well as building types such as the proportion of buildings built before the 90s and the semi-subterranean distribution.

The project was designed to resolve these conflicts by selecting sites where the amount of rainwater is the highest, the velocity is rapid, and there are many socially vulnerable groups.

서울을 기준으로 10년간의 강수량과 그에 대한 취약성을 조사한 결과, 이상 기후로 인해 지역별, 계절별 양극화는 심화되고 있었다. 그리고 그에 대한 대응 능력에는 지형, 기후와 같은 타고난 것 이외에 노년층 비율, 기초수급자 비율과 같은 사회적 지표만이 아니라 90년대 이전에 지어진 건물 분포 비율, 반지하 분포와 같은 건축 유형 또한 영향을 주고 있음을 확인했다.

빗물의 양이 가장 많고 그 속도가 급진적이며, 사회적으로 비에 대한 피해에 취약한 계층이 많이 분포하는 곳을 사이트로 선정하며 그 갈등을 해소할 수 있도록 프로젝트를 진행했다.

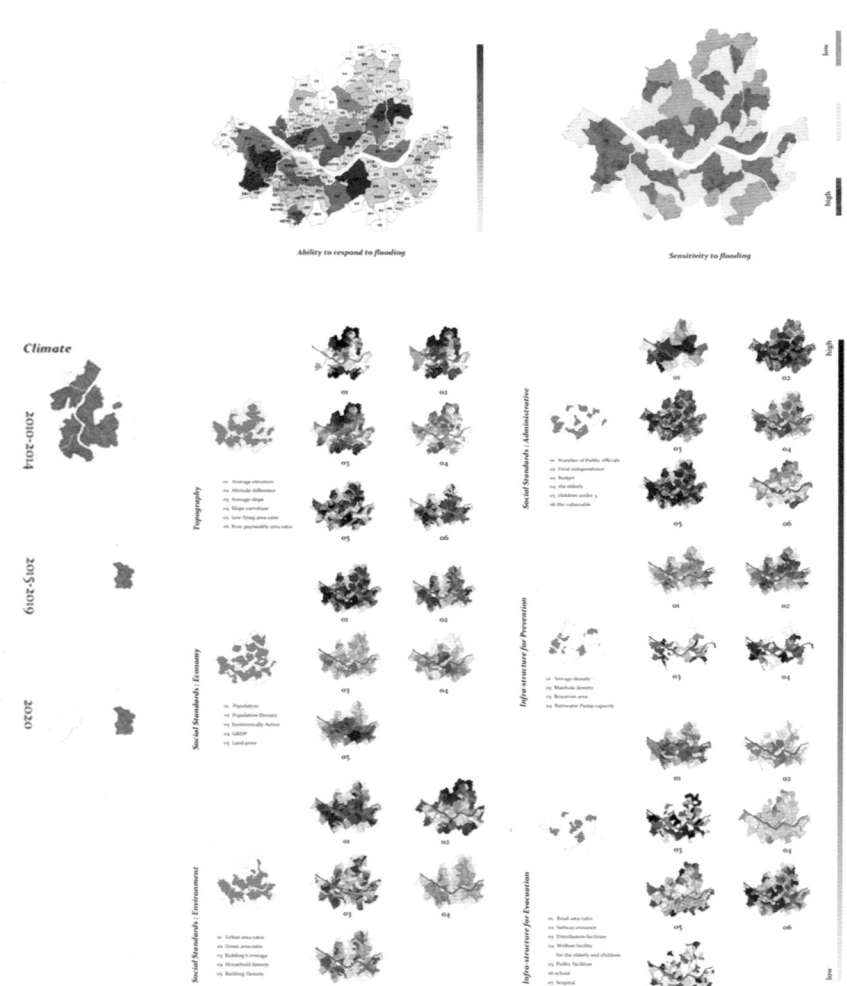

Site + Program — Seojin Kim

Beon-dong, Gangbuk-gu, is the first place where permanent rental apartments for socially vulnerable people were introduced since the 90s. The same people have been living there ever since, working, getting married, and raising children. For residents who cannot leave the neighborhood due to physical and economic conditions, plants can bring a new sensation to their daily lives.

An examination of Seoul's precipitation data reveals that abnormal weather has been causing heavy rainfall in the north of the city, and Beon-dong's steep slopes and location in the valleys of two mountains make rainwater run off more quickly. In a neighborhood with two main types of housing, there is a need to introduce architectural measures to slow water.

강북구의 번동은 90년대 이후 사회적 취약계층을 위한 영구 임대 아파트가 최초로 도입된 곳이다. 입주 이후 지금까지 동일한 사람들이 살아가고 있으며, 그들은 번동에서 직장을 얻고, 결혼을 하고 자녀들도 키우며 살아간다. 물리적, 경제적 여건으로 번동을 떠날 수 없는 주민들의 일상에 식물은 새로운 감각을 부여할 수 있다.

서울의 강수량 데이터를 조사한 결과, 이상 기후로 인해 강북에 집중적인 폭우가 이어져 왔고, 번동의 가파른 경사와 두 산의 골에 위치한 번동의 입지는 빗물을 더욱 빠르게 만들었다. 크게 두 가지로 나누어지는 주거 유형이 반복되는 번동에, 느린 물을 만들 수 있는 건축적 도입이 필요하다.

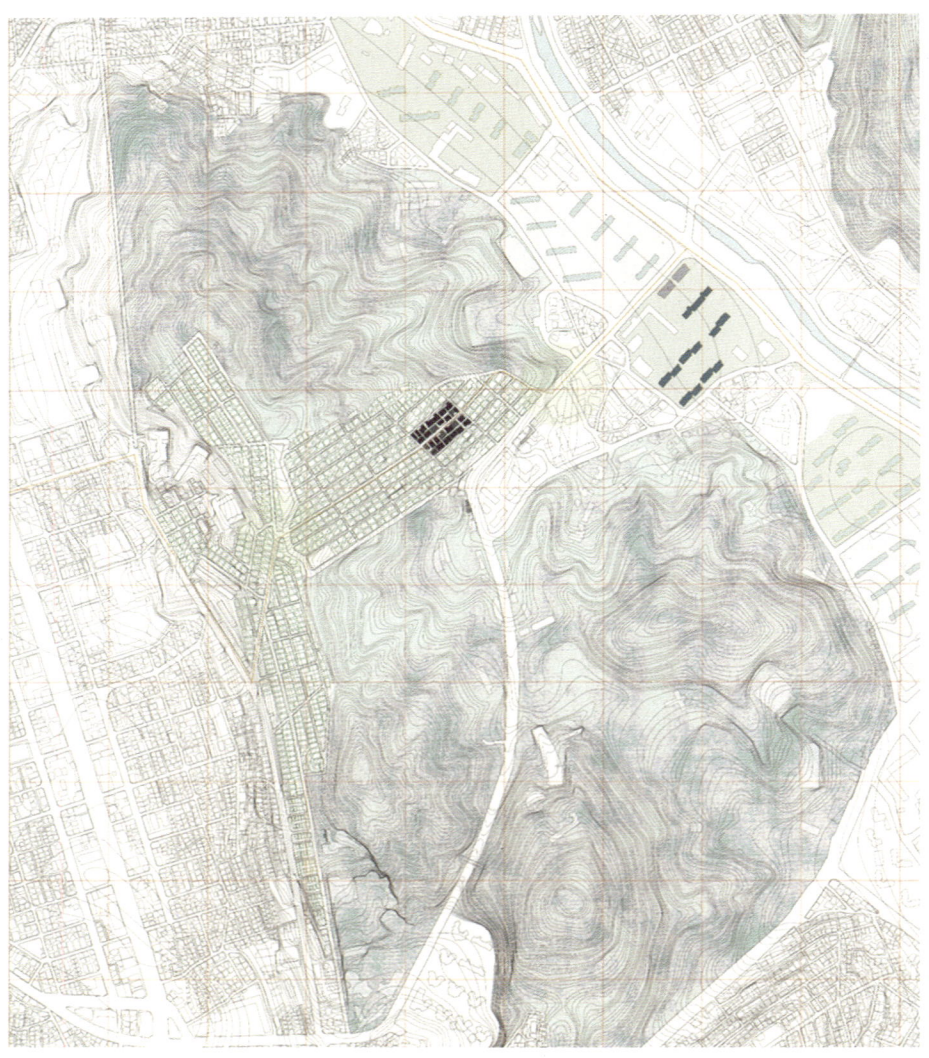

Rain Festival

As the second generation has moved in after the first, new problems are emerging, and several programs need to be added. Public activities centered around rainwater and plants are needed, such as learning spaces for young children, communal laundry facilities, and large spaces and communal gardens where people can enjoy the rain together on rainy days. Since the apartments were once used for shared childcare and joint labor, the program was designed to create a living space where residents could support each other. They also considered the influx of outsiders by creating a bathhouse that not only the residents but also the surrounding neighbors wanted.

Also, the scale of the public space is floor by floor and unit by unit (three floors). It was planned by scale so that it could be used by outsiders, and the route was planned so that it could be connected to welfare facilities for the elderly and disabled who needed direct assistance. The ramp from the roof to the ground becomes a slow-flowing path for rainwater and a meeting place where people can connect vertically.

번동에 처음 자리 잡은 사람들 이후 2세대가 나타나며 새로운 문제가 대두되고 있다는 점에서 여러 프로그램이 추가될 필요가 있다. 빗물과 식물을 매개로 공공 활동이 이루어지는 것을 중심으로 어린 학생들을 위한 학습 공간이나 공동 세탁 시설, 그리고 비가 오는 날 함께 비를 즐길 수 있는 대공간과 공동 텃밭을 마련한다. 과거 공동 육아와 공동 노동을 함께 했던 아파트였기 때문에 서로를 도우며 살아갈 수 있는 주거 공간이 되도록 프로그램을 계획했다. 또한 주민들뿐만 아니라 주변의 이웃들도 원했던 목욕탕을 만듦으로써 외부인의 유입 또한 고려했다.

또한 공공 공간의 규모는 층별, 유닛별 (세 개 층별). 그리고 외부인도 함께 사용할 수 있도록 규모별 계획이 이루어졌고 또한 직접적인 도움이 필요한 노년층 및 장애인을 위한 복지시설과의 연계가 가능하도록 동선을 계획했다. 지붕에서 지상까지 연결되는 램프는 빗물이 느리게 흐르는 길이 되면서 동시에 수직적으로 연결될 수 있는 만남의 장소가 되기도 한다.

The two building types are based on the same principle, but in different forms. They are differentiated by the way they respond to rainwater according to their location and terrain.

In the case of apartments, the post-and-beam structure is built with false walls to accommodate the changing number of family members. With this in mind, the existing units were divided into two types, one for 1-2 people and one for 3 or more people, so that similar types could share public spaces, and the renovation was divided into a north façade facing the corridor and a south façade facing each terrace. The north façade faces the existing corridor and connects to a long corridor that runs from the roof to the ground, creating a public space. The south façade extends each unit's terrace to create a corridor terrace that connects each floor while allowing for individual plantings. This creates a floor-by-floor community and allows for interaction.

두 가지 건축 유형은 같은 원리이지만 다른 형태를 지녀야 한다. 각각의 입지와 지형에 맞춰 빗물에 대응하는 방식이 형태적으로 구분된다.

아파트의 경우 '영구 임대 주거'로 이후 가족 구성원 수가 변화할 것을 감안하여 기둥-보 구조에 가벽을 세운 형태이다. 이 점을 통해 기존의 유닛을 1-2인 가구와 3인 이상 가구 두 가지 유형으로 나누어 유사한 유형끼리 공공 공간을 공유하도록 했고, 복도와 접하는 북쪽 파사드와 각 테라스와 접하는 남쪽 파사드로 나누어 리노베이션을 진행했다. 북쪽 파사드는 기존의 복도와 접하면서 지붕부터 지상까지 이어지는 긴 복도와 연결되며 공공 공간을 만든다. 남쪽 파사드는 각 유닛의 테라스를 확장해 개인의 식물을 키울 수 있으면서 각 층별로 연결되는 복도 테라스를 만든다. 층별 커뮤니티를 만들며 서로 간의 왕래도 가능하다.

The multi-family housing type implements community through block-by-block planning. They are located on slopes and have narrow spacing between buildings. The difference in level allows rainwater to flow safely and utilize some of it. Unlike apartments, they are less amenable to expansion, so approach them as components that can play a similar role. In conjunction with the existing building's water system, small additions can be made to collect and channel rainwater while also allowing for personalized planting. By changing the architectural elements such as windows, walls, and roofs, the relationship between rainwater and the building can be changed to maintain the same function as the apartment type. Both types of rainwater systems work in their respective locations as similar rainwater systems, but with the nature of water flowing from high to low, Bun-dong works as a system and transforms into a community that uses rainwater to grow plants.

다세대 주택 유형은 블럭 단위의 계획을 통해 공동체를 구현한다. 경사지에 위치해 있으며 건물 간의 간격이 좁다. 레벨 차를 이용해 빗물을 안전하게 흘려 보내고 그 일부를 이용한다. 아파트와 달리, 증축이 어려운 환경임을 감안하여 유사한 역할을 할 수 있는 요소(component)로서 접근한다. 기존 건축물의 우수 시스템과 연계하여 약간의 덧붙임을 통해 빗물을 모으고 흘려 보내는 것과 동시에 개인의 식물을 가꿀 수 있다. 창문, 담, 지붕 등 건축적인 요소들의 변화를 통한 빗물과 건축의 관계의 변화를 만들어내어 아파트 유형과 같은 기능을 유지한다. 두 유형은 유사한 빗물 시스템으로 각각의 위치에서 작용하지만 높은 곳에서 아래로 흐르는 물의 특성과 함께 번동은 하나의 시스템으로 작용하며 빗물을 이용해 식물을 키우는 공동체로서 변모한다.

Site Plan — Seojin Kim

The existing apartments are equipped with welfare facilities in each complex for close and direct care. The expansion is centered on a 10-story apartment building that includes welfare and commercial facilities, and a 15-story apartment building that is exclusively residential.

The parking lot on the ground floor will be eliminated and a communal garden will be operated. The existing small-scale garden was expanded to allow residents to grow and share crops together. Along with the interior spaces at ground level, semi-exterior spaces that share shade are also arranged along the building.

A program to increase the influx of outsiders is also planned for the lower floors. Bathhouses and commercial establishments will be located there, and ramps leading up to the roof will allow them to enjoy the rain on the roof.

The southern and northern façades provide waterways and space for plants. A three-story bridge directly connects the welfare and residential areas and allows for active interaction. Inside, the plants maintain the temperature and humidity of the interior and serve as a gathering space for residents on each floor. In the past, corridors were the only shared space for residential units, but now they have been expanded to include spaces for various activities. Activities between 1-2 person household units and 3-4 person household units are different in many ways, and joint programs are proposed accordingly. The scale of width and height is organized around the elderly, who are mainly in wheelchairs.

For the multi-family housing type, the height difference of the existing buildings was maximized to determine the direction and storage of rainwater. Most of the water collected from the multi-family houses flows quickly and safely to the lower floors of the bustling neighborhood. The water is also connected to the ground floor of the apartments, allowing for slow and controlled use of as much water as possible.

기존의 아파트는 가까이서 직접적으로 돌볼 수 있도록 단지 마다 복지 시설을 갖추고 있다. 복지시설 및 상업 시설을 포함하는 10층 규모의 아파트와 주거로만 이루어진 15층 규모의 아파트를 중심으로 증축을 계획했다.

지상층의 주차장을 없애고 공동 텃밭을 운영한다. 기존에 있던 작은 규모의 텃밭을 확장되어 주민들이 함께 농작물을 기르고 나눈다. 지상 레벨의 내부 공간과 함께 그늘을 공유하는 반외부 공간도 건물을 따라 배치된다.

또한 외부인의 유입을 늘리는 프로그램도 저층부에 계획된다. 목욕탕이나 상업 시설이 배치되고 지붕까지 이어지는 램프를 통해 그들 또한 옥상에서 비를 즐길 수 있다.

남쪽과 북쪽의 파사드는 물길이자 식물을 키우는 공간을 제공한다. 3층 높이에 배치된 다리는 복지시설과 주거 공간을 직접적으로 연결하고 적극적인 교류가 가능하다. 내부의 식물들은 실내의 온도와 습도를 유지하고 층별 주민 간의 모임 공간으로 이용되기도 한다. 과거에는 주거 유닛의 확보를 위해 복도만이 유일한 공유 공간이었다면 이제는 다양한 활동을 할 수 있는 공간으로 확장되어 사용된다. 1-2인 가구 유닛끼리의 활동과 3-4인 가구 유닛의 활동은 여러 면에서 차이가 있어 그에 맞춰 공동 프로그램을 제안한다. 주로 휠체어를 타는 노령층을 중심으로 폭과 높이에 대한 스케일을 구성했다.

다세대 주택 유형의 경우 기존 건물이 가지는 높이 차를 최대한 이용하여 빗물의 방향과 저장 방식을 결정했다. 다세대 주택에서 모은 대부분의 물은 빠르고 안전하게 번동의 저층부로 흘러간다. 그 물 또한 아파트의 지상층과 연결되는 시스템을 갖추어 최대한 많은 물을 천천히 이용하고 조절할 수 있도록 한다.

Plan + Section — Seojin Kim

Perspective Drawing — Seojin Kim

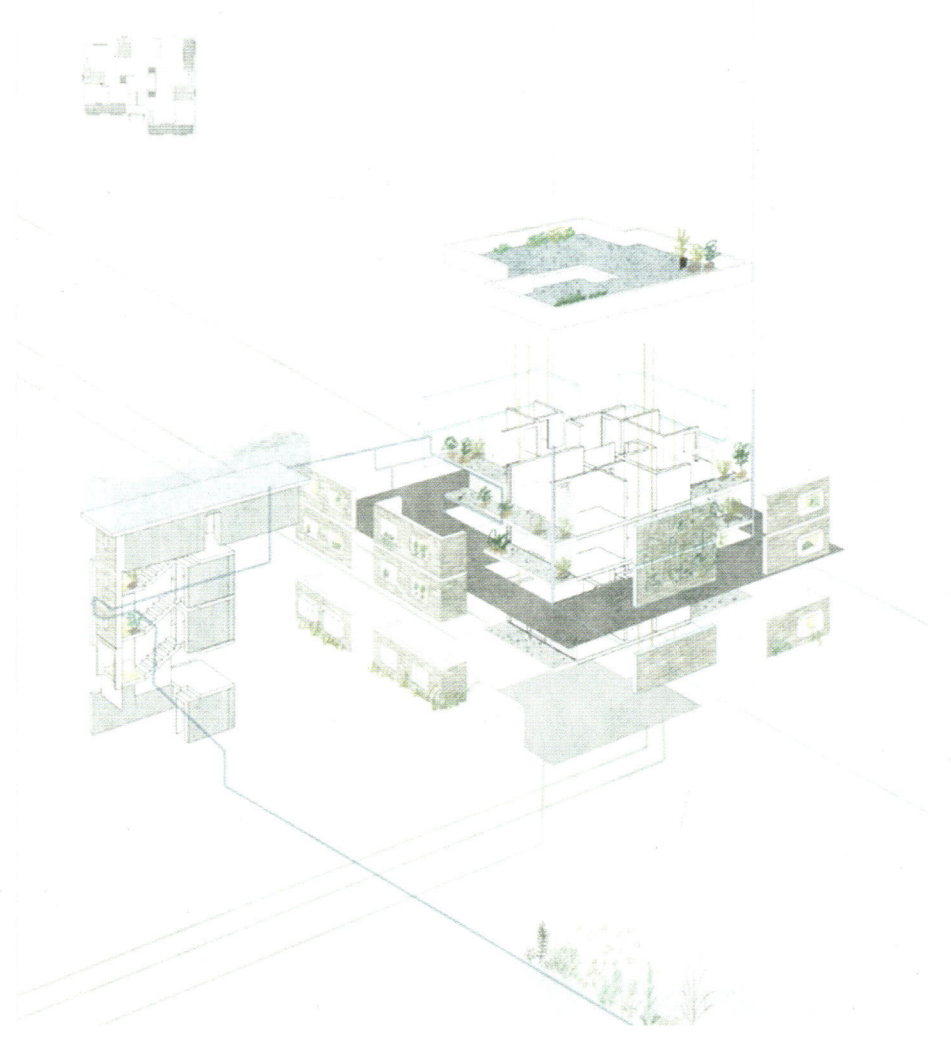

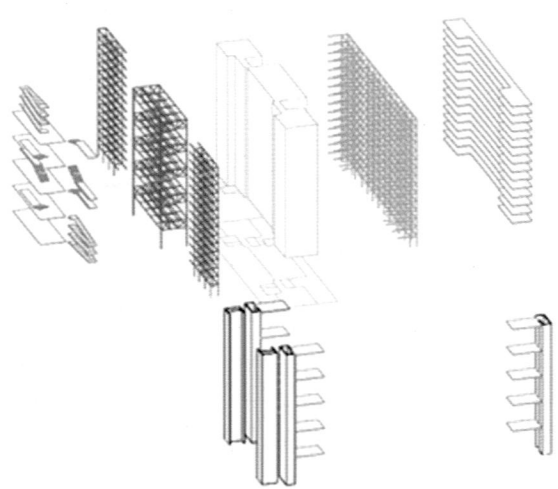

The post-and-beam structure of the existing building was identified, and a new structure was built with a safe separation distance from it. The new structure has similar column spacing to the existing building and also supports a new roof that collects water. Due to the age of the existing apartments, the new structure was designed to act as an independent structure rather than a structural connection.

The north façade of the building creates a public space shared by all three floors. It is cantilevered using trusses to extend between the existing building cores and serves as a public space.

The rainwater is divided into two categories: exposed water and water flowing through pipes. Exposed rainwater is directed to flow slowly through the public spaces and water the plants. By creating slow water, it also affects the humidity and temperature inside. Some water needs to be interchangeable or stored quickly, so pipes are used to deliver it. In each case, they are partially connected to the existing water storage system.

In addition, the water system used in the past in the existing apartment complex was renovated and used as a rainwater reservoir, and it was planned to work as a system that actively receives and manages the amount of water by connecting with the city's water supply system and nearby rivers.

기존 건물이 가지는 기둥-보 구조를 파악하고, 그로부터 안전을 위한 이격 거리를 가지고 새 구조물을 증축했다. 새로운 구조는 기존 건물과 유사한 기둥 간격을 가지면서 물을 모으는 새로운 지붕을 지지하는 구조물이기도 하다. 또한 기존의 아파트가 오래되었기 때문에 구조적인 연결보다는 독립적인 구조체로서 역할을 하도록 했다.

건물의 북쪽 파사드의 경우 3개 층이 함께 사용하는 공공 공간이 생겨난다. 트러스를 이용한 캔틸레버 형식으로 기존 건물의 코어 사이에 뻗어 나와 공공 공간으로 사용된다.

빗물은 크게 외부로 노출된 물과 파이프관을 통해 흐르는 물이 있다. 외부로 노출된 빗물은 공공 공간을 통과하고 식물에 물을 대주며 느리게 흐르도록 유도 되었다. 느린 물을 만들어 내부의 습도와 온도에 영향을 주기도 한다. 일부 물은 서로 교환이 가능하거나 빠르게 저장되어야 하기 때문에 파이프관을 이용해 물을 전달한다. 각각의 상황을 고려하여 기존의 우수시스템과 일부 연결되어 사용된다.

또한 기존 아파트 단지에서 과거부터 사용하던 우수 시설을 보수하여 빗물 저장고로 사용하고 도시 내의 급수 시설과 근처의 하천과도 연계하여 물의 양을 적극적으로 수용 및 관리하는 하나의 시스템으로서 작동하도록 계획했다.

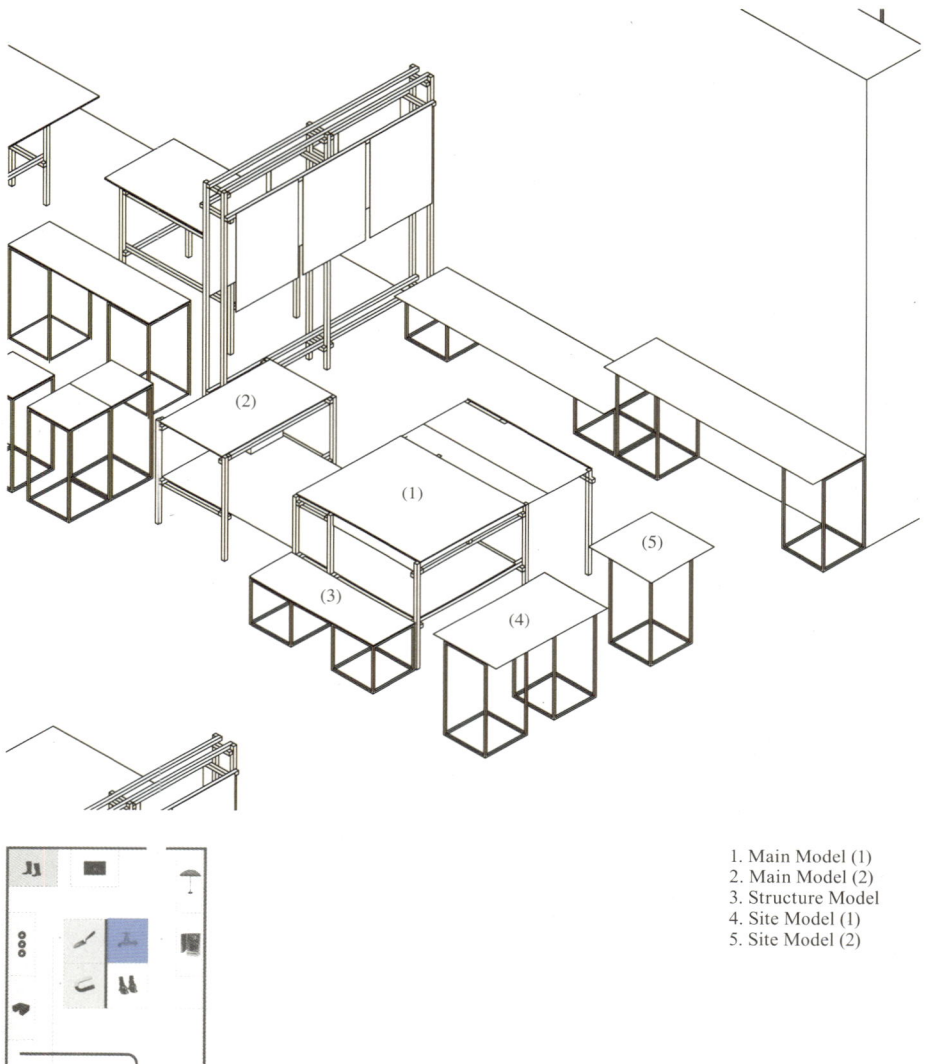

1. Main Model (1)
2. Main Model (2)
3. Structure Model
4. Site Model (1)
5. Site Model (2)

졸업도우미　허해인
　　　　　　양진모. 윤감수. 김연경. 허진아.
　　　　　　한지원.

MAIN MODEL

Scale 1:100
3D print PLA + Royal board, 60 by 120 cm, 60 by 120 cm

SITE MODEL

Scale 1:2500
3D print, 60 by 60 cm

Scale 1:1000
Paper, 60 by 120 cm

STRUCTURE MODEL

Scale 1:200
Royal board, 60 by 60 cm

The Mounds, Artifacts　　Taeyoung Kim
　　　　　　　　　　　　　Studio

　　　　　　　　　　　　　신지승
　　　　　　　　　　　　　Jiseung Shin

The Mounds, Artifacts

The exhibition itself is an extension of the work. The theme of 'land,' which was important in the work, was revealed in the horizontal plane, which was also emphasized in the exhibition furniture. It was a long table measuring 2500mm by 6000mm. Since it was attached to the wall by 6000mm, the goal was to make the viewer face the wall directly and linger for a long time without glancing at it as they walked by.

Long horizontal plates were spaced at odd intervals to form a floating display desk, and an equally long horizontal panel was placed on the wall. There was a protruding fire hydrant located 4500mm away from the left end. I covered it with all the work I had done in school except for my graduation project, my fifth grade notebooks laid on the stall. And I leaned my conceptual model against, and taped the title of the project on a black sheet of paper. I turned the flaws of the space into an integral part of the exhibition. I made a display desk by myself for six pieces measuring 1000mm by 1400mm to emphasize the horizontal plane.

The paper portfolio is a time-consuming object to look at. Also, there are no chairs in the exhibition hall, so I put two chairs at the right end that are not higher than the horizontal surface of the desks but have a backrest so that you can sit for a long time. Also, I inserted a stand light that partially divides the space with the stalls and helps you see the portfolio.

It feels weird to graduate. During the semester, whatever happened, I could just soak myself in the works in the school workroom. But now, I feel like I need to put some distance between myself and the company I work for, think long term, and start taking up hobbies and exercising. It feels both awkward and comfortable at the same time to spend time away from my work. Six years seem like a blink of an eye. I miss it, and also there are many times when I don't think about it.

전시하는 것 자체가 작업의 연장이다. 작업에서 중요했던 화두 '땅'이, 전시 가구에서도 강조돼 보이는 수평면에서 드러나도록 했다. 2500×6000의 가로로 긴 자리였다. 6000만큼 벽에 붙어있는 자리였기 때문에, 관람객이 지나가다 흘겨보지 않고, 벽을 정면으로 마주하게 하며, 오래 머물게 하는 것이 목표였다.

긴 수평 판이 묘한 간격을 가지고, 떠있는 형태의, 전시대를 놓고, 벽에 똑같이 가로로 긴 판넬을 두었다. 왼쪽 끝에서 4500 떨어진 지점에 돌출된 소화전이 있었다. 이를 졸업 작업을 제외한 학교에서 했던 모든 작업과, 5학년 작업 노트, 그리고 개념 모델을 기대어 놓을 수 있는 가판대로 가리고, 그 위에는 검은색 시트로 작업 제목을 뽑아 붙였다. 이 자리의 결점을, 전시의 필수적인 요소로 전환했다. 1000×1400의 책상 6개를 수평면을 강조할 수 있는 형태로 전시대를 직접 제작했다.

실물 포트폴리오는 보는 데에 시간이 오래 소요되는 대상이다. 또한 전시장에는 의자가 없다. 따라서 오른쪽 끝에 책상의 수평면보다 높지 않지만, 등받이가 있어 오래 앉아있을 수 있는 의자 두 개와, 가판대와 함께 공간을 일부 분절해주며 포트폴리오를 보는 데에 도움이 되도록 하는 스탠드 조명을 두었다.

졸업하니 기분이 이상하다. 어떤 일이 생기든 학기 중에는 학교 생활이 몸을 그냥 담가 놓으면 됐었는데, 지금은 내가 속한 회사와 적당한 거리도 두어야 하고, 장기적인 생각도 해야하고, 취미생활과 운동을 시작해야만 할 것 같은 기분이 든다. 내 작업을 안 하는 시간을 보내고 있는 것이 어색하기도 편하기도 하다. 6년이 덧없게 느껴지기도 하고, 그립기도 하고, 아무 생각이 안 들 때도 많다.

166   The Mounds, Artifacts

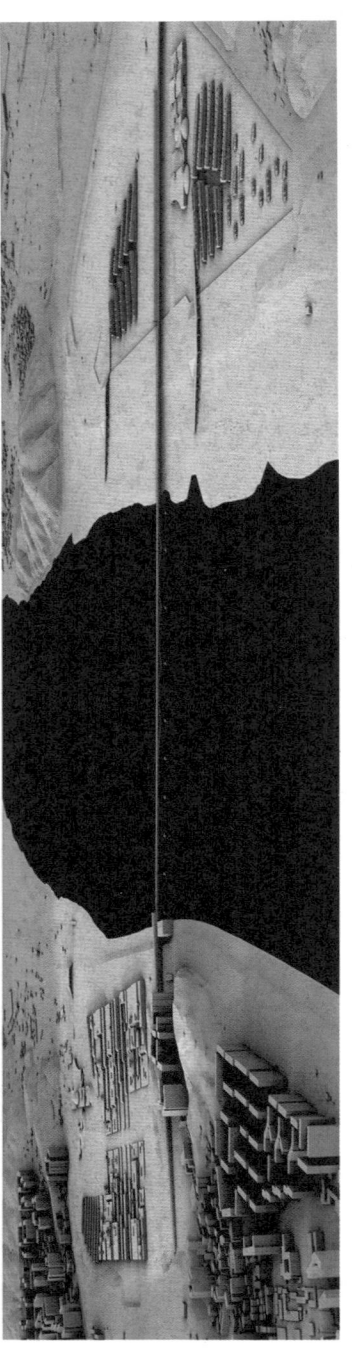

I think I have to live my life by figuring out what I am and where I am, and I worked on a space for such a life. In the artificial mountain of folds, the useless class lives. Fundamentally, we are the beings without meaning, beings who work and see, beings existing by the Earth.

The majority will become the irrelevant class, the meaningless people. It is only when we move away from production and labor that we can consider what the fundamental act is. In order to remain human, they will seek to live on a spectrum from the active to the contemplative, from materializing production to meditating by doing nothing. Assuming that a survival is a solid future, a housing will be subordinated to this space of activity and contemplation.

For this life, they go to a land that has been discontinuously emerged by technological development. On top of the modernized and undergrounded water treatment center, two huge but different plots of land, facing each other across the river, are folded. Several mounds on each site create an artificial fold from the waterfront to the city, depicting the circularity of life between the two.

The design has a sequence and a process. Voids that were partially excluded from the city, are laid out first, followed by buildings, such as spaces for working and seeing (ac-tivity-contemplation), which were also pushed out from a city, and then residential areas, and so on, in the reverse order of many plans.

Those who doubt their foundation, will live on, in, and through the artificial mound, where the experience of the earth becomes the norm. Soil will be excavated and covered over the top of the mound; underground facilities will exist simulta-neously with a reconstructed interior and a mountainous surface above; even new buildings will be mounds.

내가 무엇인지 어디 있는지 알아가면서 살아야 한다고 생각한다. 그런 삶을 위한 공간을 작업했다. 주름의 인공산에서 무용계급이 거주한다. 우리는 본질적으로 의미가 부여되지 않은 존재이며, 작업하고 들여다보는 존재이며, 땅에 의한 존재다.

다수는, 의미 없는 존재인 무용계급이 될 것이다. 생산과 노동에서 멀어졌을 때 비로소 근본적인 행위가 무엇인지 생각해본다. 이들은 인간으로 남기 위해 활동적 삶부터 관조적 삶까지의 스펙트럼 (물질로 실체화하는 제작부터 아무것도 하지 않는 명상까지)에서 살고자 하게 된다. 생존을 확보된 미래로 가정하고, 이 활동-관조의 공간에 주거 공간이 종속된다.

이 삶을 위해, 기술 발전에 의해 불연속적으로 떠오른 땅으로 간다. 현대화, 지하화된 물재생센터 상부에 생긴 거대한 같지만 다른 두 대지는, 강을 사이에 두고 서로 마주보는 주름진 곳이다. 각 대지에 여러 마운드들이 강변부터 도시까지를 걸친 인공 주름이 되고, 두 곳을 오가는 순환적인 삶을 그린다.

설계는 순서와 프로세스가 있다. 도시에서 일부 배제되었던 보이드, 작업하기와 들여다보기(활동-관조)를 위한 공간이 먼저 놓이고, 그 다음 건물, 주거 영역 등이 이전에 배치된 것들의 영향 아래 들어서는 식으로, 다수의 계획안과 역순이다.

본인의 기반을 의심하는 이들은, 땅의 경험이 일상이 되게하는 인공 마운드 위에서, 동시에 그 속에서, 거주할 것이다. 흙은 파내지고 마운드 상부에 덮어질 것이고, 지하 시설 위에는 개축된 내부공간과, 상부의 산 같은 표면이 동시에 존재할 것이며, 신축된 건물들조차 마운드가 될 것이다.

1. 허구와 실제, 기계혁명과 노동, 무용계급, 보편기본소득,
   자아, 진정성, 장식의 모호함, 공과 사, 또 다른 지구 리서치

2. 물재생센터 사이트 리서치

3. 본질적인 삶의 스펙트럼

4. 사이트에 건축적인 대응

5. 공간 스터디

6. 리뷰

Through the research, I start by thinking about what kind of a person is and what kind of life the person needs to live. Then I sketch out an architectural vision of that life, how the person should respond to it, etc. Find a site for the person. When I arrive at the site, I figure out what kind of land it is. And then, I apply the hypotheses I've developed from the information I've gathered to the site. Lastly, I abstract the life and space to the land. After weaving the context, I design the building.

I study the people, the relationships, the life, the quality of the experience of the inner space, not just the bird's eye view. I'm not an engineer, neither a decision maker, rather I'm an architectural designer. So the spaces I design, I try to make them look like the work of an architecture student who doesn't rely on a specific discipline. Then I look back to see if the proposal and the space are valid. When I do this, I conclude that it's definitely 'visionary' work, but it has power and value for it's a reference to a 'solution'.

리서치를 통해 어떤 사람이, 어떤 삶을 살아야 하는지부터 생각했다. 그 다음 해당 삶을 위한 건축적 상상, 대응 방식 등을 대략적으로 그린다. 그를 위한 사이트를 찾아나선다. 사이트에 도착하면, 어떤 땅인지 파악한다. 지금까지 세운 가설과, 수집한 정보와 추상적으로 그렸던 삶과 공간을 이 사이트에 적용한다. 맥락을 직조하고나서, 건물을 설계한다.

조감도에서 그치지 않는, 사람과, 관계와, 삶과, 내부 공간 경험의 질을 위한 스터디를 한다. 이때 나는 엔지니어도, 결정권자도 아닌, 건축 설계를 하는 사람이다. 그러니 내가 설계한 공간은, 특정 분야에 의지하지 않는 건축 전공자의 작업으로 보이고자 했다. 그 다음 이 제안과 공간이 유효한지 돌아본다. 이 작업을 할 때, 분명히 visionary하지만, solution에 참고가 된다는 점에서 힘과 가치가 있다고 정리했다.

170

The Mounds, Artifacts

When you think you have become irrelevant on this planet, you might want to go to another planet. What would you do when you see you and others useless and meaningless? Where should humans look forward to with their lives? How will society change, and what will an architecture for such conditions look like? So far, we've been looking at how humans have supported their lives. Especially, a fiction has been a means of personal identity and social co-operation, allowing for labor, production, and growth.

본인이 이 지구에서 무관한 존재가 되었다고 생각할 때, 또 다른 행성에 갔다 오고 싶을 것이다. 본인과 타자가 무의미하다고 생각될 때 어떻게 해야 하는가? 인간은 본인의 삶을 어디에 기대야 하는가? 사회는 어떻게 변할 것이며, 그에 따른 건축은 어떠할 것인가? 지금까지 인간들은 본인의 삶을 지탱하는 방식을 들여다보고자 했다. 허구는 개인의 정체성, 사회적 협력의 수단이었고, 이로 인해 노동, 생산, 성장이 가능했다.

The Mounds, Artifacts

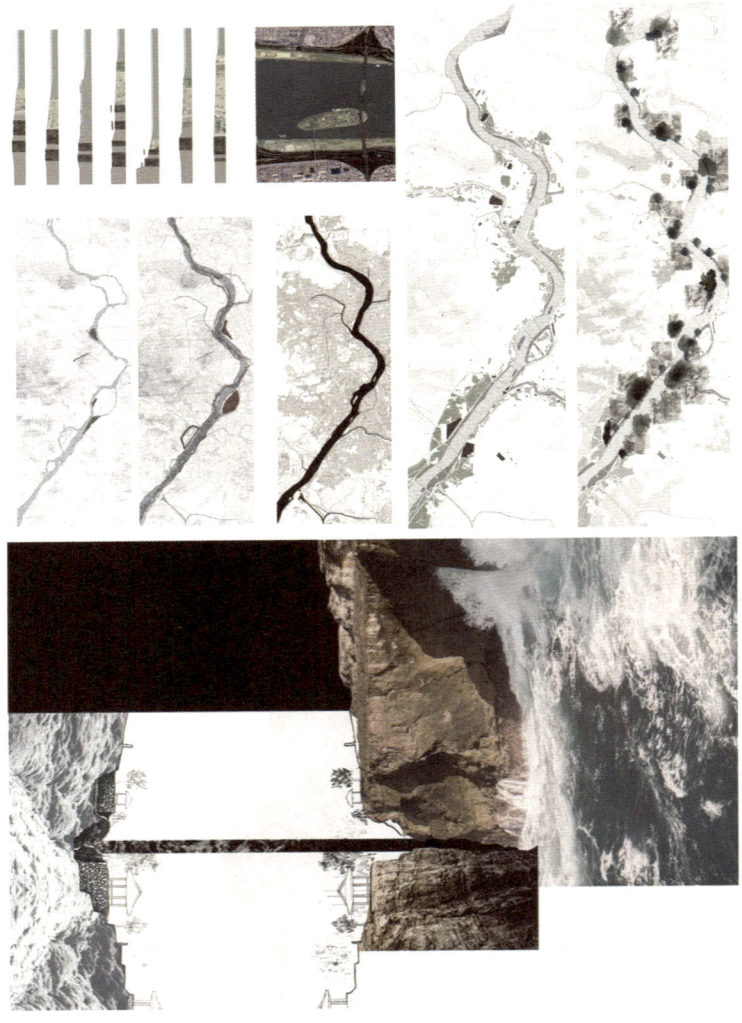

The flooding of machines makes us think about the essence of human life. Where should the useless class, which is still emerging little by little, expect the value of life beyond survival without labor? While thinking that the free will and authenticity are all fictions, I would like to ask about the fundamental human life outside of labor and capital.

How could we deal with publicness? I researched how collectives and architecture work. At Soseowon, a space for temporary housing, exchange, rest, and hiding in nature, I saw two spaces facing each other and a stream between them. The relationship between the two structures, defined by water and topography, is a synesthetic experience.

기계의 폭주로, 인간 삶의 본질이 무엇인지 생각하게 된다. 지금도 조금씩 생기고 있는 무용계급은 노동 없는 생존을 넘어서 어디에 삶의 가치를 기대야하나? 자유의지도, 진정성도 모두 허구라는 생각이 드는 와중에, 노동과 자본 밖에서의 근본적인 인간의 삶에 대해 질문하고자 한다.

공공성을 어떻게 풀어갈 것인가. 집단과 건축의 작동 방식을 리서치했다. 임시 주거, 교류, 쉼, 자연 속 은거의 공간인 소쇄원에서, 마주보는 두 공간과, 그 사이의 냇물을 보았다. 물과 지형으로 정의되는 두 건축물의 관계와, 공감각적 경험이다.

The Mounds, Artifacts

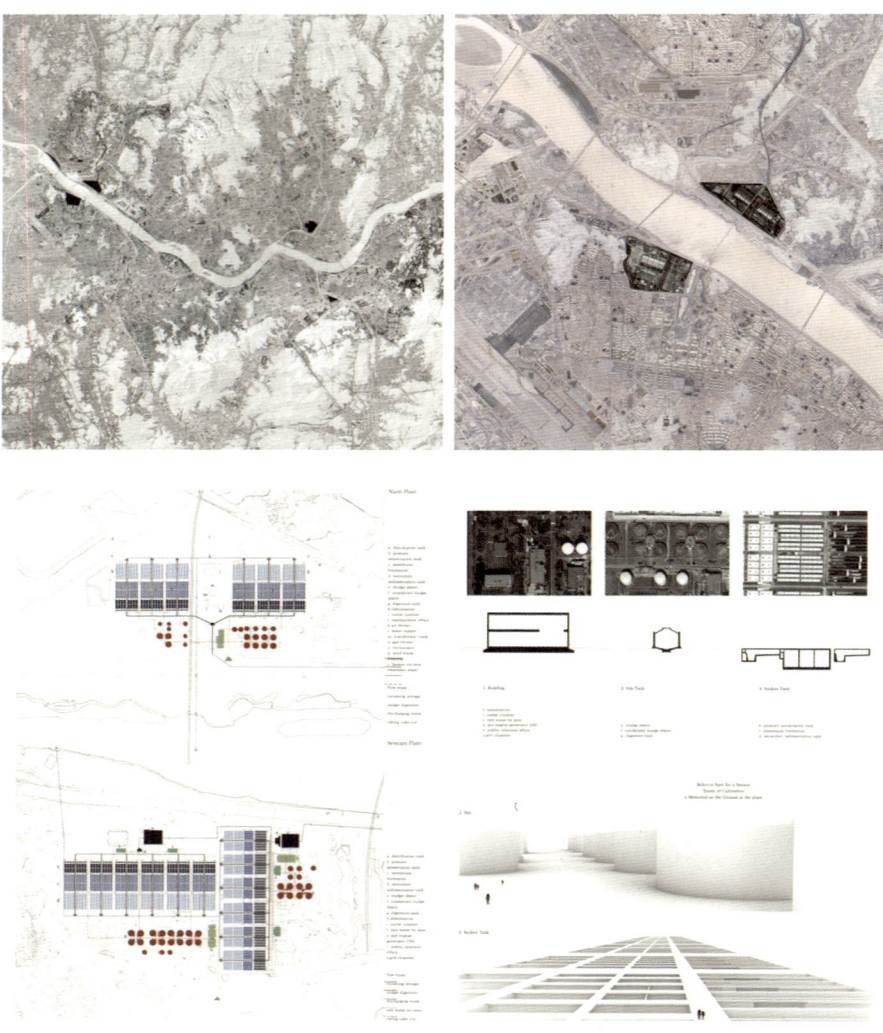

Site + Program — Jiseung Shin

Emerging Land

An empty space on the edge of the city, just outside the city center, is perfect. This is the Seonam and Nanji Water Treatment Center, one of four sewage treatment plants on the border of Seoul. It is to be modernized, reduced in volume, and buried underground. There will be a large, discontinuous expanse of empty land on top. I want to deal with the discontinuity through appropriate severence. There are times when memories should not be forgotten, preserved, but there are also times when they should be forgotten. This new land is a crack created by advancing technology and changing times. The spaces of the meaningless class by the advancement of technology and the changing era will land here.

Opposing Land

Two similar facilities face each other across a river. It is a condition that makes us constantly dream of another earth, see others, and see ourselves. Moving back and forth between the two facilities, the people of the useless class live a life that is not static. They reflect on where they have been and look forward to where they are going.

떠오른 땅

도심에서 약간 벗어난 도시 경계부의 빈 공간이 적합하다. 서울의 경계부에 네 개의 하수처리장 중 서남, 난지 물재생센터다. 부피가 줄어든 채, 현대화되어 지하화될 예정이다. 상부에 불연속적으로 거대한 빈 땅이 생긴다. 적절한 단절을 통한 불연속성을 다루고자 한다. 기억이 잊히지 않아야 하고, 보존되어야 하는 경우도 있지만, 망각해야 할 때도 있기 때문이다. 이 새로운 땅은 기술이 발전하고 시대가 변화하여 생긴 틈이다. 기술이 발전하고 시대가 변화하여 생긴 무용계급을 위한 공간이 이 땅에 들어선다.

마주보는 땅

비슷한 두 시설이 강을 사이에 두고 마주보고 있다. 끊임없이 또 다른 지구를 꿈꾸게 하고, 타자를 바라보게 하고, 본인을 바라보게 하는 조건이다. 두 시설을 오가면서, 무용계급의 사람들은 정체되지 않는 삶을 산다. 지나온 곳을 성찰함과 동시에 앞으로 갈 곳을 전망한다.

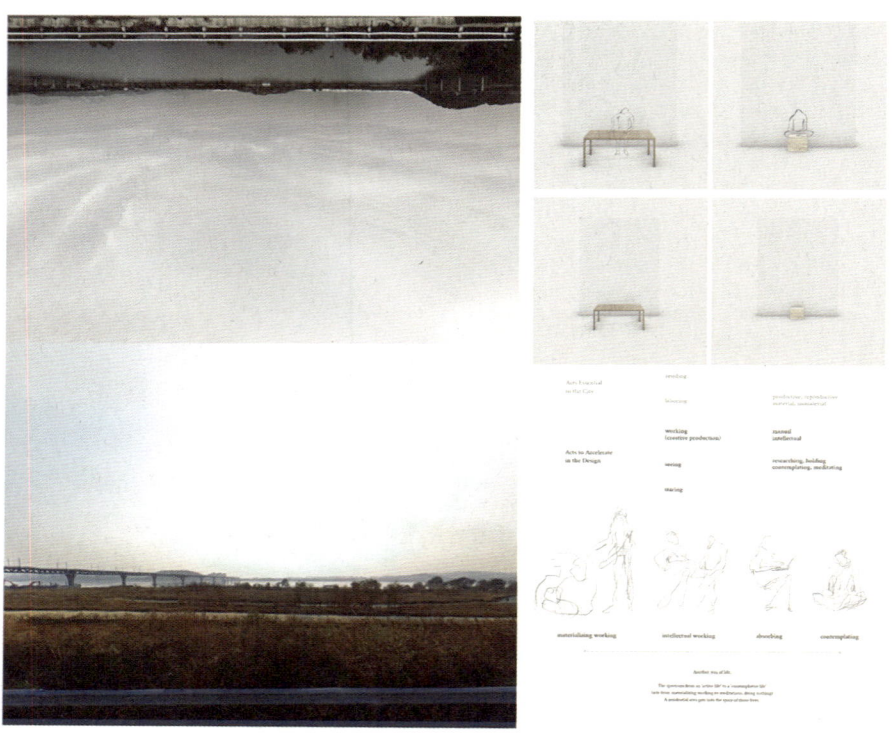

Site + Program — Jiseung Shin

Active Life — Contemplative Life

A life that ranges from materializing creation to doing nothing but meditation. It presupposes an essential activities beyond human survival and labor. These are: working through physical materialization, working with immaterial things, understanding and studying what is given, reflecting, and meditating without thinking. These programs are called 'working' and 'seeing.' Not everyone has the opportunity to fully engage in these activities. Especially since those acts are often excluded by labor. If machines take over labor and production, it would be meaningful to design spaces for these essential acts. A residential space would be subordinate to the space for these activities, as opposed to the current dichotomous separation of housing and work sites.

활동적 삶 — 관조적 삶

물질로 실체화하는 제작부터 아무것도 하지 않는 명상까지의 삶이다. 인간의 생존과 노동을 넘어선 본질적인 활동을 전제한다. 물리적 실체화를 통한 작업하기, 비물질적인 작업하기, 주어진 것을 이해하고 공부하기, 성찰하기, 아무 생각도 하지 않고 명상하기 등이다. 이 프로그램들을 '작업하기'와 '들여다보기'로 정리한다. 이 활동들을 온전히 할 수 있는 기회가 누구에게나 주어지지 않는다. 특히 노동에 의해 배제되는 경우가 많기 때문이다. 기계가 노동과 생산을 대신한다면, 사람이 행위들을 위한 공간을 설계하는 것이 의미 있을 것이다. 이 공간은, 현재의 주거와 노동 현장의 이분법적인 분리와 반대로, 이 활동을 위한 공간에, 주거가 종속되어 있는 형태를 가진다.

산으로 건축적 대응을 하고자 했다.
땅의 연장이지만 지형 그 자체는 아닌
인공 지형으로서 말이다.

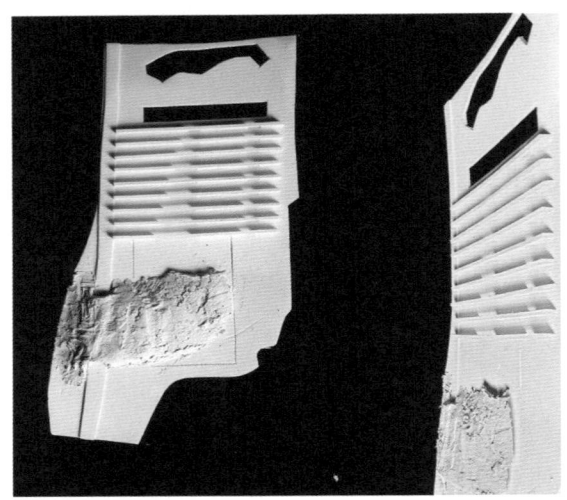

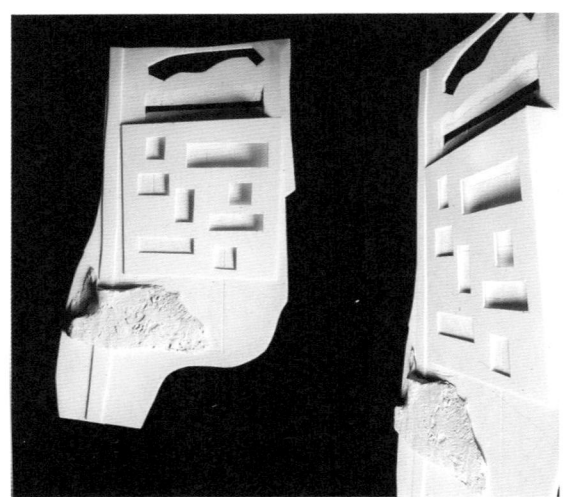

I wanted to create an architectural action with the mountains. As an artificial landscape, which is an extension of the land, not just a terrain.

As for the architectural proposal, the order reveals its value. The void, the mound as a carpet, is laid first. This carpet will largely be the exterior space of the landscape, the loose program. This reversal of the traditional approach which is placing voids in the space left by the building, or holes in the mass of the building, differs from the seemingly rigid urban structure around it.

There is a common element in the north and south. Two elements are in common: the mound as a topographical experience of the waterfront, cut off from the city, and the mound as a sandbox for those who feel insecure about their foundation, to touch and transform the ground with their hands.

On the north side, the carpets in different strips are laid horizontally first, and a long wall mound crosses the carpets vertically. The spectrum of life, consisting of four programs, runs parallel to the building's orientation. Toward the center, the building rises, an act of reflection, and outward, an act of work.

On the south side, a rectangular void is placed first. The boundaries of the four programs span the void, and surrounding the void is a series of tiered podium mounds with wall mounds that break up the mass. Toward the center, it gets lower, noisier, and more complex.

Four programs and dwellings traverse parallel to the building orientation. Different types of mounds at each site will create the same but different conditions and skies.

건축적으로 대응하는 방식은, 순서에서 그 가치가 드러난다. 보이드, 카펫으로서의 마운드를 먼저 놓는 것이다. 이 카펫은 대체로 조경, 느슨한 프로그램의 외부 공간이 될 것이다. 건물을 놓고 남는 공간에 보이드를 두거나, 건물 매스를 두고 구멍을 뚫는 기존의 방식과 역순이라는 점에서 주변의 경직돼 보이는 도시 구조와 차이가 있다.

북쪽과 남쪽에 공통된 요소가 있다. 도시와 단절된 강변을 지형적인 경험으로 넘어다닐 수 있게 하는 마운드와, 기반이 불안하다고 느끼는 자들이 땅을 제 손으로 만지고 변형시키는 경험을 하는 샌드박스로서의 마운드다.

북쪽은 서로 다른 띠 형태의 카펫들을 가로로 먼저 놓고, 세로로 긴 벽 마운드가 카펫들을 횡단한다. 4개의 프로그램으로 이루어진 삶의 스펙트럼이 건물의 방향과 평행하게 놓인다. 가운데로 갈수록 건물이 높아지고, 성찰의 행위가 몰리며, 바깥으로 갈수록 작업의 행위가 몰린다.

남쪽은, 장방형 보이드를 먼저 놓는다. 4개의 프로그램의 경계가 보이드에 걸치고, 보이드를 둘러싸면서 매스가 분절된, 벽 마운드가 집적된 여러 단의 포디움 마운드가 놓인다. 가운데로 갈수록 낮아지고, 시끄럽고 복작복작하다.

4개의 프로그램과 주거가 건물 방향과 평행하게 횡단한다. 각 사이트에 여러 종류의 마운드들이 같지만 다른 조건과 하늘을 만들 것이다.

The Mounds, Artifacts

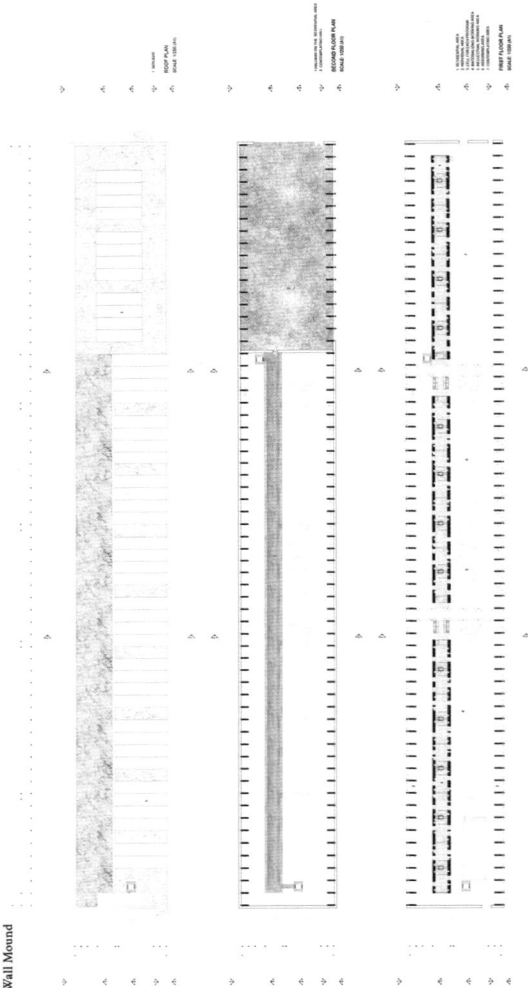

Wall Mound

Wall Mound

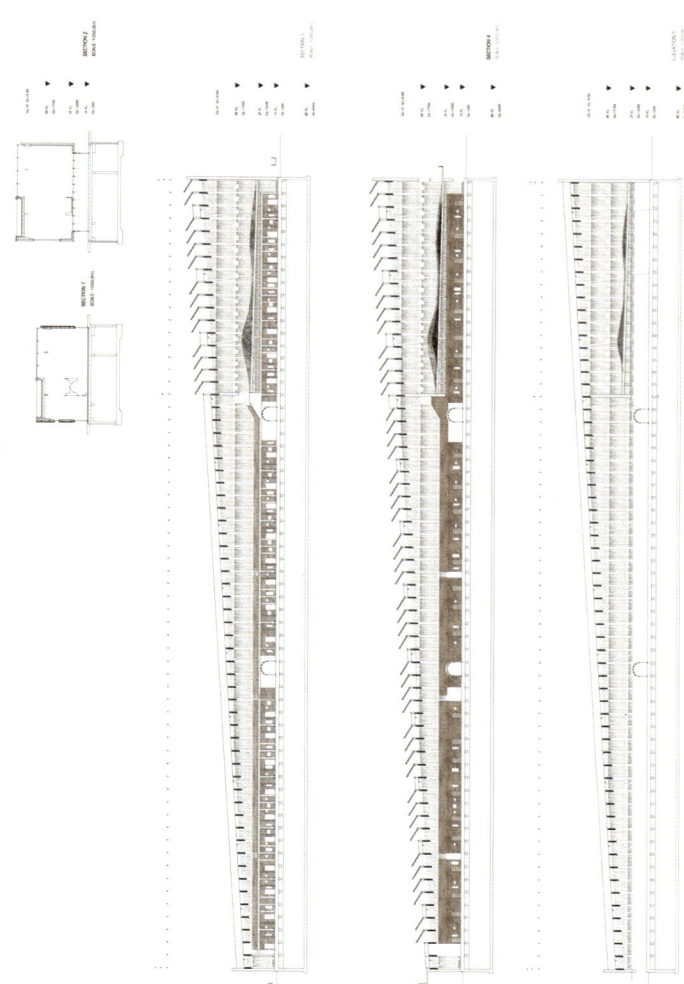

Perspective Drawing — Jiseung Shin

A wall mound is a concrete structure in the basement with a wooden top. The structure is made up of steel plates sandwiched between CLTs and bolted together. There is a brace every four spaces. Inside is an earthen living space. The roof details, with rooftop landscaping, are as follows: building slab, waterproofing layer, protective mortar, poured concrete, drainage board, non-woven fabric, drainage soil, nurturing soil, topsoil, irrigation layer, wood chips, and drainage drains.

벽 마운드는 지하의 콘크리트 구조 위에 상부는 목조다. CLT 사이에 철판을 끼우고 볼트로 고정하는 구조다. 4칸에 한 번씩 가새가 들어간다. 내부에 흙구조의 주거 공간이 들어간다. 옥상 조경에 의해 지붕 디테일은, 건축 슬라브, 방수층, 보호 몰탈, 누름 콘크리트, 배수판, 부직포, 배수용 인공토, 육성용 인공토, 표토, 관수 장치층, 우드 칩, 배수 드레인 순으로 이루어져 있다.

The Mounds, Artifacts

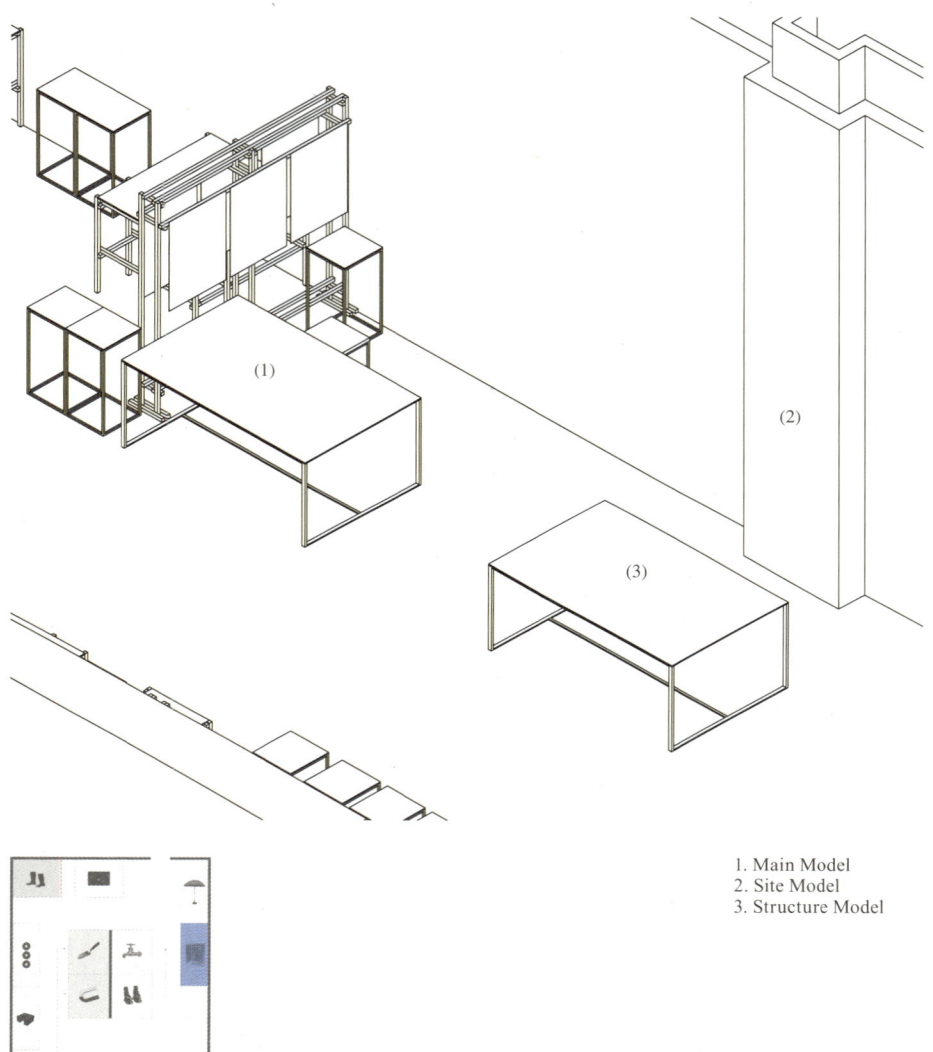

1. Main Model
2. Site Model
3. Structure Model

졸업도우미

이지민
강다현. 김가은. 김의하. 이승헌. 이지헌.
이한결. 정선아. 정선혜. 정지우. 조혁재.

MAIN MODEL

Scale 1:100
Royal board + wood + clay, 20 by 128 cm

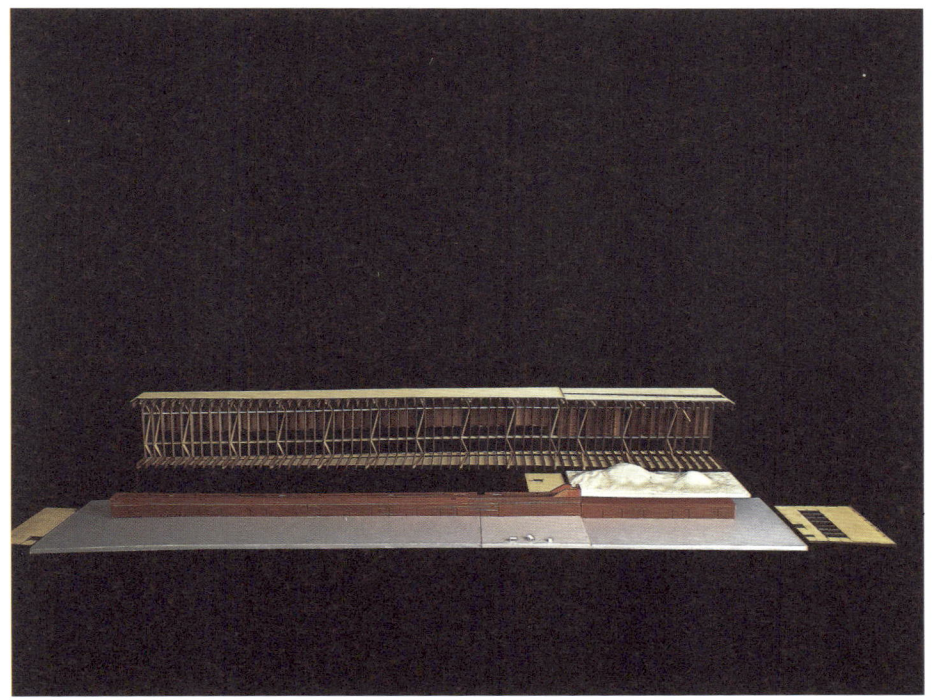

SITE MODEL

Scale 1:2000
Royal board + 3d print PLA, 200 by 140 cm

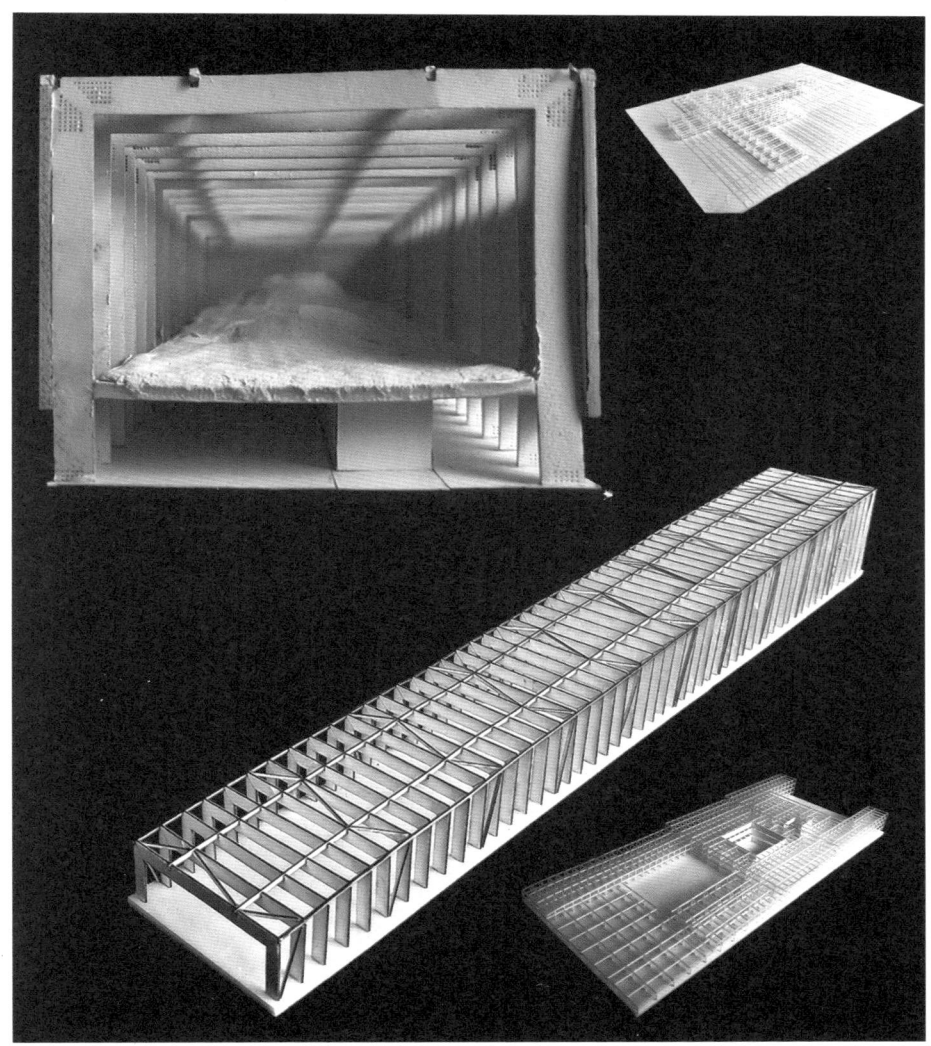

STRUCTURE MODEL

Scale 1:200
Royal board, 10 by 64 cm

Dongmyo: *Intergenerational Network*	Taeyoung Kim Studio
이연우 Yonu Lee	

The exhibition space was divided into three areas:

Part 1, where you can see the models of each prototype (Anchor, Passage, and Arcade);

Part 2, where you can see the site model at a glance when the prototypes were placed; you can see the portfolio and picture frames that explain the entire project;

Part 3, where the video of Dongmyo Market is being played;

Part 3, where you can experience the atmosphere of Dongmyo Market more vividly while watching the video by placing simple tables and chairs that are commonly used in Dongmyo Market.

전시 공간은 총 세 개의 공간으로 나누었다. 각 프로토타입(Anchor, Passage, Arcade) 모델을 만나볼 수 있는 파트 1.

그리고 그 프로토타입들이 배치되었을 때의 모습을 한눈에 볼 수 있는 사이트 모델과 전체 프로젝트를 설명해 줄 포트폴리오 및 액자들이 배치된 파트 2.

동묘시장의 생생한 모습을 담은 영상이 재생되고 있는 파트 3 까지.

파트 3 은 동묘시장에서 주로 쓰이는 간이테이블과 간이의자를 배치해 영상을 감상하는 동안 그곳의 분위기를 더 생생히 경험할 수 있도록 만들었다.

Throughout my academic career, I have been interested in boundaries. Sometimes it was between groups, sometimes between individuals. My interest in the invisible boundaries between generations as we enter a super-aged society was sparked by a chance visit to Dongmyo Market. In this particular place, I saw the possibility of breaking down the boundaries between generations, as it was almost the only place in the city where the elderly were the main users, at least in my experience.

Unfortunately, Dongmyo Market, like many other places in Seoul, seemed to have begun to gentrify as a result of media coverage. As I was worried that Dongmyo Market would lose its unique color, I started to look for ways to systematize and revitalize the market to preserve its value. In fact, during the year-long project, many shops were torn down and buildings unrelated to the old market started to go up, and I saw the market changing every time I visited. In the end, I was driven by the hope that my work would not remain as a paper project.

본인은 학창 시절 내내 경계에 관해 관심을 가져왔다. 그것은 때론 그룹 간의 경계, 개인 간의 경계이기도 했다. 초고령화 사회에 접어들면서 갖게 된 세대 간 보이지 않는 경계에 대한 관심은 우연히 방문한 동묘시장에서 촉발되었다. 이 특별한 장소에서 세대 간의 경계가 허물어질 수 있는 가능성을 보게 되는데, 이는 적어도 본인이 경험한 범위에서는 노인들이 주된 사용자가 되는 장소 중 여러 세대가 섞이는 거의 유일한 장소였기 때문이다.

하지만 안타깝게도 동묘시장 역시 서울의 다른 장소들과 마찬가지로 매스컴을 타면서부터 젠트리피케이션이 진행되기 시작한 것으로 보였다. 이대로 가면 동묘시장만의 고유한 색이 사라질 수 있다는 불안감이 일면서 동묘시장의 가치를 보존하기 위한 시장의 체계화 및 활성화 방안이 무엇이 있을까 모색하기 시작했다. 실제로 프로젝트를 진행하는 1년이라는 기간 동안 수많은 가게가 허물어지고 구제시장과는 상관없는 빌딩들이 올라서기 시작해 답사를 갈 때마다 변해 있는 시장의 모습을 보게 되었다. 결국 본인의 작업이 프로젝트로만 머물지 않았으면 하는 희망 사항이 이 작업을 끌고 간 추동력이었다.

1. Interest
2. Today's issue
3. Searching a place with resilient boundary
4. Focused on three places with specific generation
5. Site selection
6. Proposal

During the first semester, I envisioned a framework of unorganized support facilities for the users of Dongmyo Market, a variable storefront system to increase the convenience of stores and street vendors, and a roof system to bring homogeneity to the chaotic environment of the market.

During the summer workshop, I studied the spacing of the roof louvers, which I introduced to make the alley a coherent environment, with a focus on thermal comfort, and in the second semester, I went further and focused on the porosity of the market and the refinement of the system.

1학기 동안은 동묘시장 사용자를 위해 체계가 전무한 지원시설의 기본 틀과 점포 및 노점상의 편의를 증대할 가변 점포 파사드 시스템, 시장의 어수선한 환경에 균질성을 부여하기 위한 지붕 시스템을 구상했다.

방학 워크숍 동안은 골목을 일관성 있는 환경으로 만들어 주기 위해 도입한 지붕 루버 간격에 대해 열 환경적 쾌적성에 주목하여 연구했으며, 2학기에 접어들어서는 여기서 더 나아가 시장의 다공화 및 시스템의 구체화에 집중했다.

Dongmyo: *Intergenerational Network*

My research began with the topic of boundaries. I first looked at how people's boundaries manifested themselves in residential settings and third places. While gated communities and shantytowns have clear, visible boundaries, elderly communities and youth communities seem to have invisible boundaries.

To understand this, I looked at Topgol Park, Jongmyo Park, and Dongmyo Market, where the elderly gather, and found that while Jongno 3-ga, where Topgol Park and Jongmyo Park are located, is a world where the peripheralized elderly do menial jobs, Dongmyo Market is an environment where the elderly and young people can blur micro-boundaries through 'second-hand.'

리서치는 경계라는 주제에서 시작되었다. 먼저 사람들의 경계가 주거환경과 제3의 장소에서 어떻게 나타나는지를 살펴봤다. 게이티드 커뮤니티와 판자촌은 뚜렷이 가시적인 경계를 가진 데에 비해 노인 공동체와 청년 공동체는 비가시적인 경계를 가지고 있는 듯했다.

그 실체를 파악하고자 노인들이 모여드는 탑골 공원, 종묘 공원, 동묘시장을 살펴본 결과, 탑골 공원과 종묘 공원이 있는 종로3가가 주변화된 노인들이 소일을 하는 세상인 데 비해 동묘시장은 노인과 청년이 '구제'를 매개로 미시적 경계를 흐릴 수 있는 환경을 가지고 있음을 알 수 있었다.

204   Dongmyo: *Intergenerational Network*

Street vendors in Dongmyo Market usually pick up their goods during the week and start selling on weekends at 3am and end at 5pm. They rent one of the few restrooms in the nearby malls and pay the electricity bills of stores that are not open on weekends to operate in front of them. Dongmyo Market has the characteristic of shrinking during the week and expanding on the weekend because these vendors gather in front of stores that close on the weekend.

They use a variety of devices that can be found in everyday life for seating, spreading things out, displaying things, and holding things. Envisioning a device that could systematize them became part of the project research.

As the proportion of street vendors varies depending on whether the store is open on weekends or not, Dongmyo Market can be categorized by the nature of the alley, such as a residential alley with a mix of stores and street vendors, a market alley full of street vendors, and an alley with a mix of both stores and street vendors, which also gives the alley a distinctive character.

동묘시장의 노점상은 주로 주중에 물건을 떼어와서 주말 새벽 3시부터 장사를 시작해 오후 5시에 장사를 끝낸다. 이들은 인근 상가의 몇 개밖에 없는 화장실을 빌려 쓰며 주말에 장사하지 않는 점포에 전기세 정도를 지불하고 그 앞에서 장사하고 있다.

동묘시장은 주중에 수축했다가 주말에 팽창하는 특성이 있는데, 이는 주말에 닫는 점포들 앞에 이 노점상들이 모이기 때문이다. 그들은 자리 잡는 용도, 물건을 펼쳐 놓는 용도, 물건을 전시하는 용도, 담아 놓는 용도로 일상에서 쉽게 볼 수 있는 다양한 장치들을 사용한다. 이들을 시스템화할 수 있는 장치의 구상이 프로젝트 리서치의 일부가 되었다.

점포의 주말 동안 영업 여부에 따라 노점의 비율이 달라지다 보니 동묘시장은 점포와 노점이 섞인 주거지 골목, 노점으로 가득한 시장 골목, 점포로 가득한 골목과 점포 노점 식당 모두가 섞인 골목 등 골목의 성격으로 분류가 가능했고 이 또한 골목에 변별력을 부여하는 계기가 되었다.

Dongmyo: *Intergenerational Network*

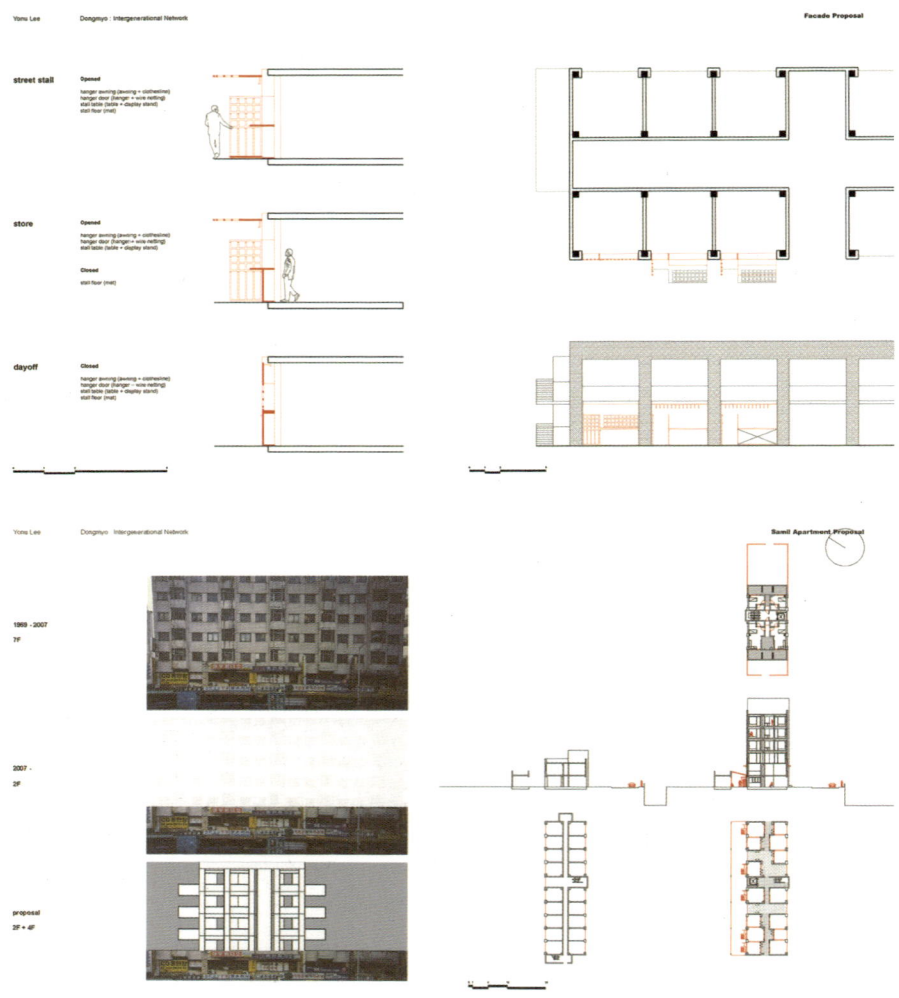

In order to revitalize this market, the residential alley needed to be connected to the main commercial space. The long alley, which is divided into three sections, seemed to be suitable for the intervention of support facilities needed by each character at each point where its character changed.

Samil Apartment, which occupies a large part of Dongmyo Market, has a peculiar history: it was maintained as a seven-story residential complex until 2007, when five floors were demolished, leaving the lower two floors of shops alone. The possibility of a vertical extension was found in the fact that the lower structure was already designed for seven floors. Therefore, a total of four floors of temporary housing for merchants were added above the existing shops.

이 시장의 활성화를 위해서 주거지 골목은 주 상업 공간과 이어질 필요가 있었다. 총 세 구역으로 나뉘는 긴 골목은 그 성격이 바뀌는 지점마다 성격별로 필요로 하는 지원시설이 개입해 들어가기에 적합해 보였다.

동묘시장의 큰 부분을 차지하고 있는 삼일아파트의 경우 2007년까지 7층의 주상복합으로 유지되다가 아래 두 층의 상가만 내버려 두고 다섯 층이 철거되었다는 특이한 역사가 있다. 하층부 구조가 이미 7층을 고려한 구조라는 점으로부터 수직 증축의 가능성을 발견할 수 있었다. 이에 따라 기존 상가 위로 총 4층의 상인들을 위한 임시주거시설을 증축했다.

Dongmyo: *Intergenerational Network*

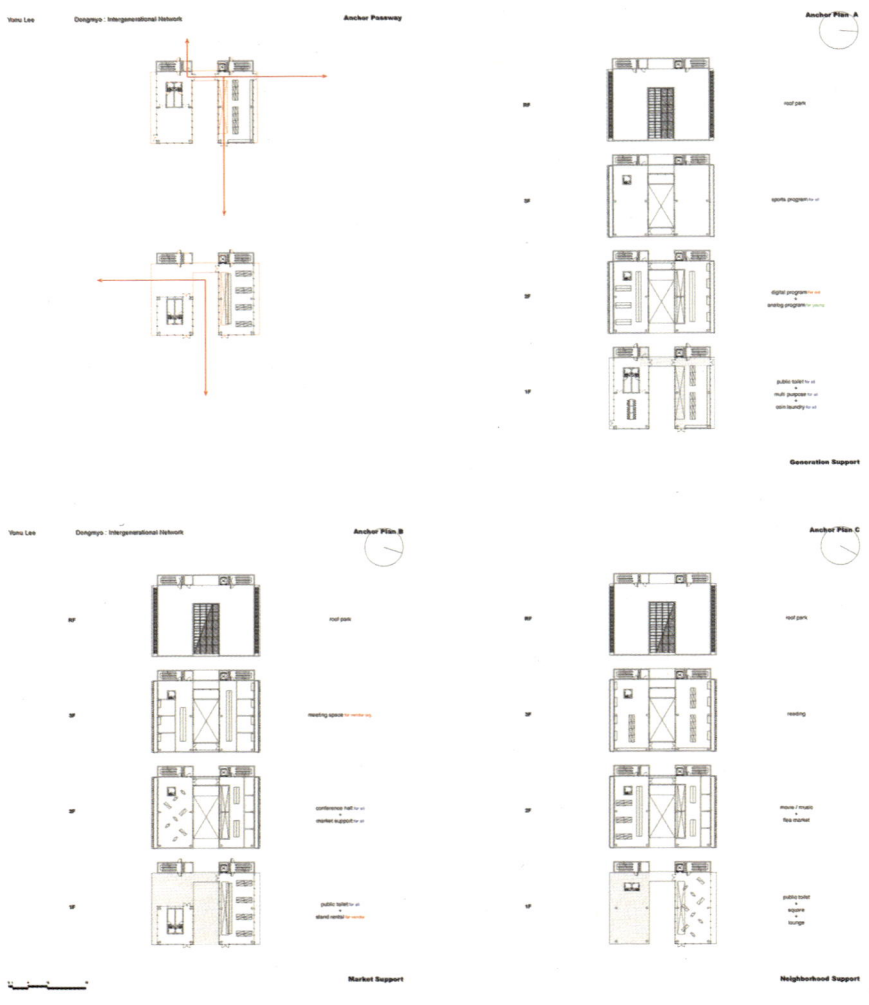

For the first and second floors, which are existing shopping centers, I demolished the existing stairwells and created an easily accessible passageway to facilitate people's access, and at the same time, I relieved a few shops to make it a passageway between shopping centers and markets. For the corridor facing the shops, I focused on expandability of the boundaries. The space can be opened or closed with false walls or folding doors, and can be expanded by acquiring the store across the street or next door.

For support facilities, the ground floor layout varies depending on where they are located: the new alley between the vocational schools is designed to allow the facility to exit through the alley to the outside, while those facing the main street have a floor plan that allows for shortcuts. Finally, for the support facility facing the market alley, the road passes underneath the facility, while providing an open space for the adjacent shopkeepers to set up their stalls and sell their items.

A functional corridor linking the residential space to the main commercial space is placed between the vocational schools, providing a space for specialization and students to practice their skills (baking or hairdressing) for free with the elderly. This is connected to a generational support facility with a similar purpose.

기존 상가인 1-2층의 경우 사람들의 원활한 접근을 위해 기존 계단실을 철거하고 쉽게 접근할 수 있는 통로를 냈으며 동시에 몇 곳의 상점을 덜어내어 이를 상가와 시장의 통로로 만들었다. 상가가 마주 보는 복도는 경계의 확장성에 중점을 뒀다. 가벽 또는 폴딩도어로 공간을 열거나 막을 수 있으며 건너편 또는 옆의 상점을 매입하는 경우 확장해서 사용할 수도 있다.

지원시설의 경우, 어디에 위치하는지에 따라 1층의 공간구성이 달라졌다. 학교 사이로 새로 낸 골목은 시설을 통해 바깥 골목으로 빠져나가게끔 설계했고 대로변에 면한 시설의 경우 지름길을 선택할 수 있는 평면으로 설계했다. 마지막으로 시장 골목에 접한 지원시설의 경우 시설 아래로 도로가 통과할 수 있도록 함과 동시에 바로 옆 상가 사람들이 매대를 늘어놓고 판을 벌일 수 있는 공터를 마련했다.

주거 공간과 주 상업 공간을 엮어주는 기능의 통로는 학교 사이에 두어 특성화고 학생들이 노인을 대상으로 무료 실습할 수 있는 공간(제빵 또는 미용)을 제공한다. 이는 유사한 취지의 세대 지원시설과 연결되어 있다.

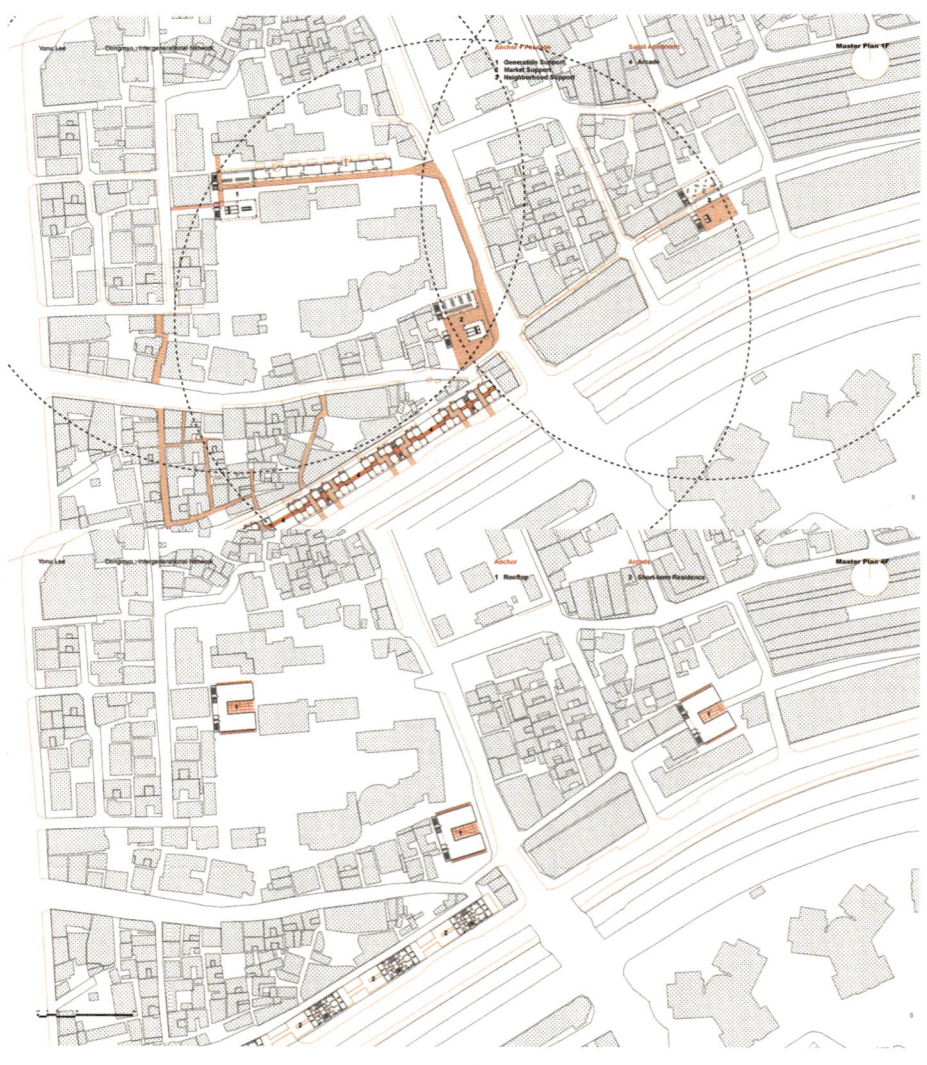

Site Plan

Yonu Lee

I placed a generation support facility to break down the boundaries between generations near the housing for young people and the school where students can be trained during their volunteer work for elderly, a market support facility to break down the boundaries between street vendors and storefronts near a cluster of shops and stalls, and a neighborhood support facility near a church and a senior center.

I also proposed to connect the passageway between the support facilities and the school and the interior and exterior floor paving of the Samil Apartment so that these spaces can be perceived as one in the chaotic marketplace.

청년주거와 노인을 대상으로 실습할 학교와 가까운 곳에는 세대 간의 경계를 허물어 줄 세대 지원시설을, 상가와 노점이 밀집한 곳에는 노점상과 점포상의 경계를 허물어 줄 시장 지원시설을, 성당과 노인정 근방에는 동네 지원시설을 배치했다.

또한 지원시설들과 학교 사이에 위치한 통로 그리고 삼일아파트의 내 외부 바닥 페이빙을 연결해 혼란스러운 시장이라는 장소에서 이 공간들이 하나로 연결되어 인식되게끔 제안하였다.

Dongmyo: *Intergenerational Network*

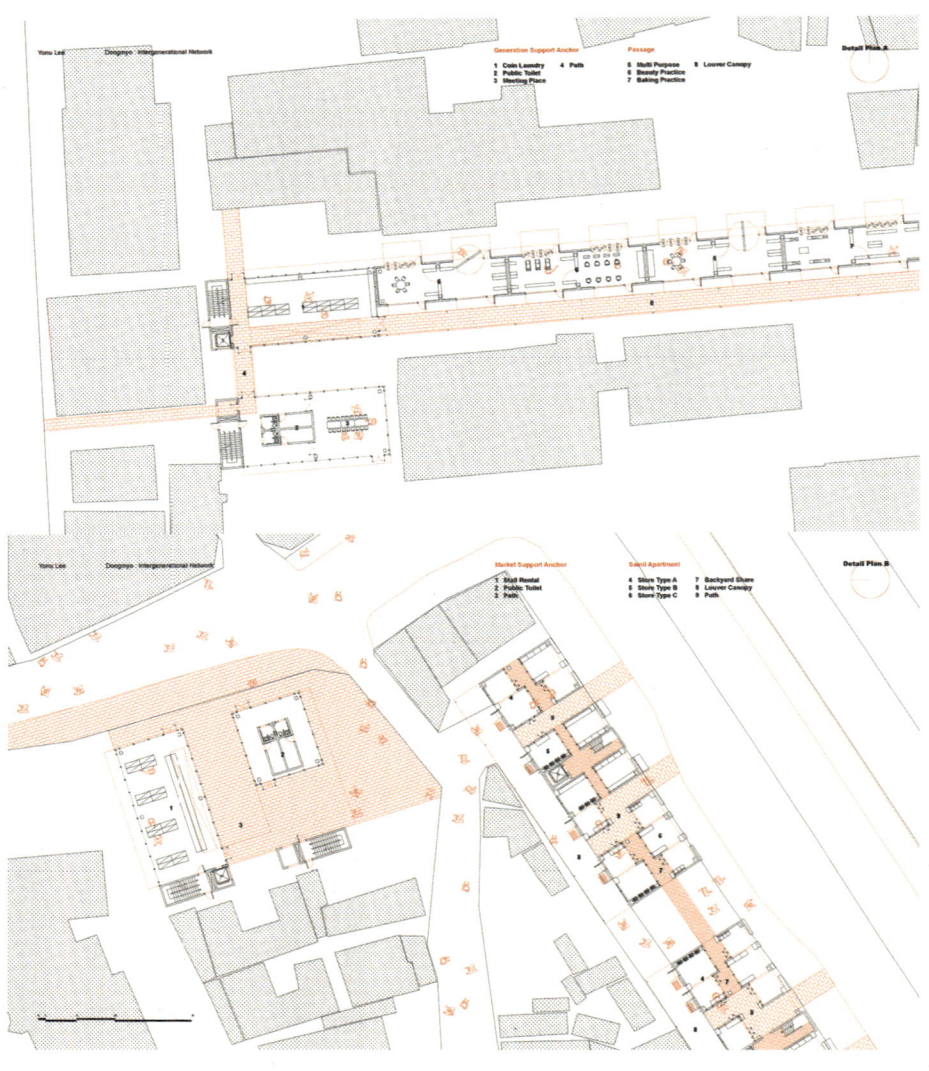

Plan + Section — Yonu Lee

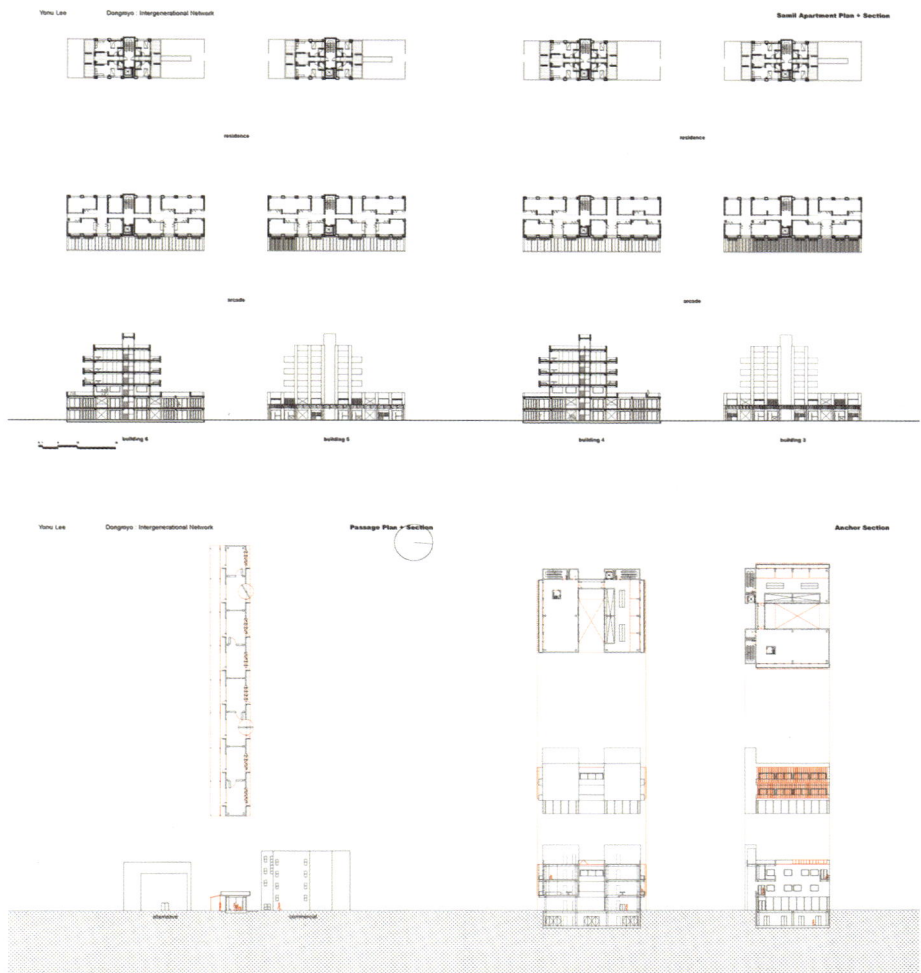

Dongmyo: *Intergenerational Network*

Perspective A : Generation Support Anchor

Perspective A : Passage

Perspective B : Arcade

Perspective B : Street

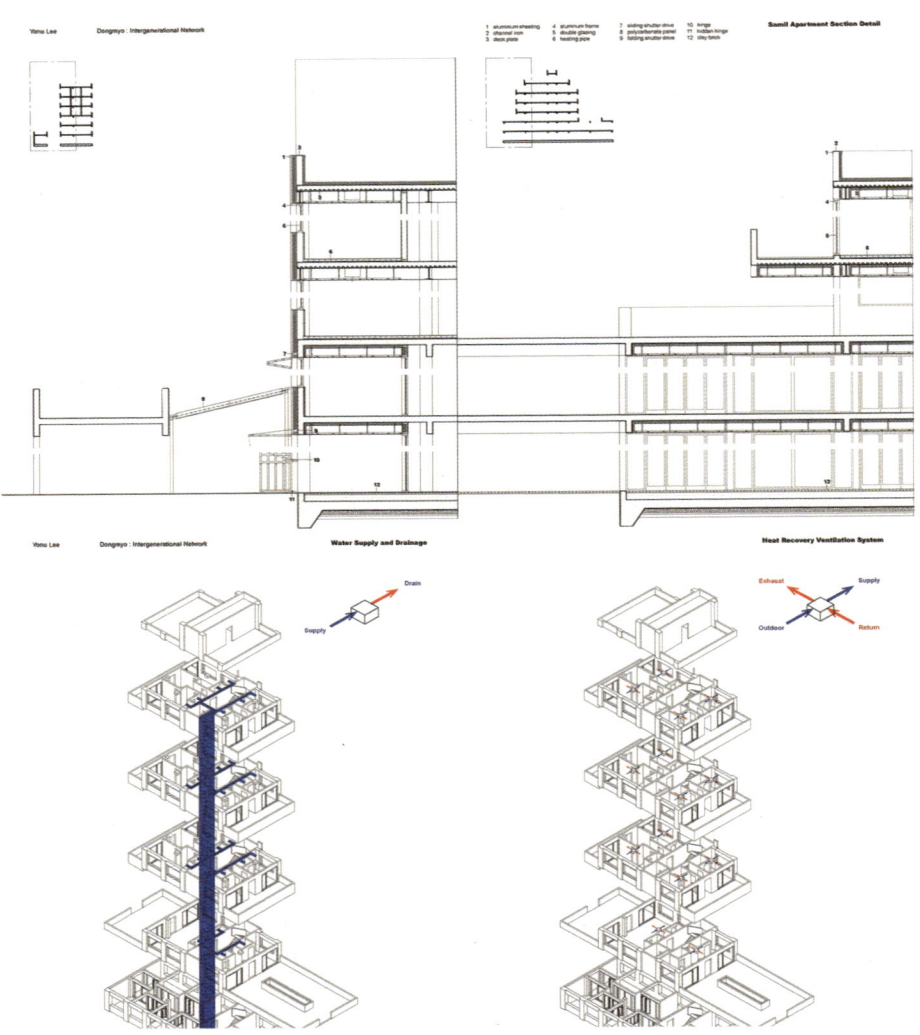

Two of the three prototypes proposed in this project are new construction and one is a remodeling and extension.

In the case of the remodeling and extension, I chose to use a steel structure for the extension to reduce the structural burden on the underlying structure. For the linear space located in the passage between the schools, I chose a system that is independent of support facilities and structures despite being attached side by side for structural stability.

본 프로젝트가 제안하는 세 개의 프로토타입 중 두 개는 신축, 한 개는 리모델링 및 증축이다.

리모델링 및 증축을 진행하는 삼일아파트의 경우 하부의 구조적 부담을 줄이기 위해 철골구조를 이용해 증축하는 방법을 택했다. 학교 사이 통로에 위치한 선형 공간의 경우, 구조적 안정성을 위해 외관상으로는 나란히 붙어있음에도 불구하고 지원시설과 구조로부터 독립하는 형식을 취했다.

Dongmyo: *Intergenerational Network*

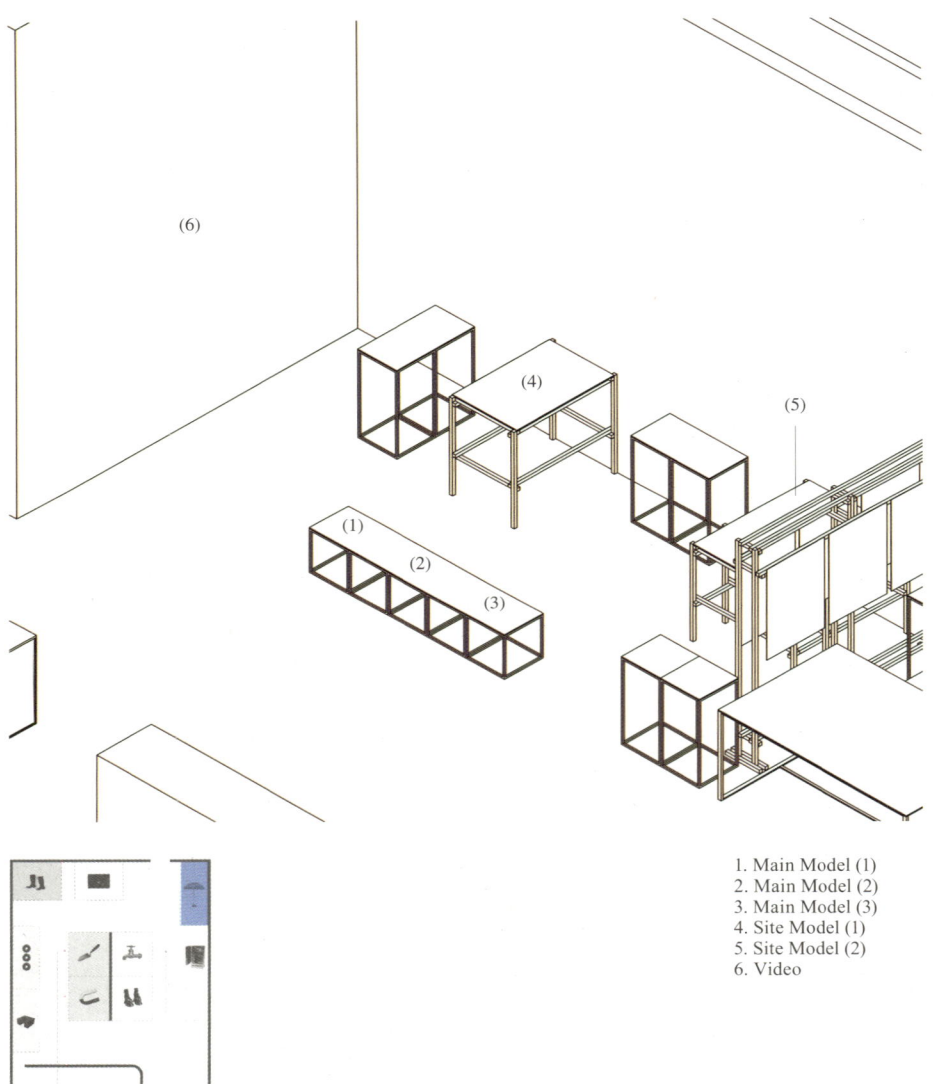

1. Main Model (1)
2. Main Model (2)
3. Main Model (3)
4. Site Model (1)
5. Site Model (2)
6. Video

졸업도우미

유경림
김민형. 고베니. 김회연. 유현욱. 윤은.
조은. 이유신

MAIN MODEL

Scale 1:100
Acrylic + Paper + 3D print, 38 by 17 by 24 cm, 30 by 24 by 18 cm, 36.5 by 14 by 5 cm

Dongmyo: *Intergenerational Network*

SITE MODEL

Scale 1:1000
3D print, 76 by 57 by 12 cm

Scale 1:400
Basewood + Acrylic + paper + 3D print, 84.1 by 84.1 by 25 cm

Model — Yonu Lee

STRUCTURE MODEL

Scale 1:100
Paper, 38 by 11 by 24 cm
Paper, 27 by 22.5 by 18 cm
Paper, 73 by 8 by 5 cm

The Fourth Attack:
Linking Instabilities

Sooyoung Kim
Studio

이은후
Eunhoo Lee

The Fourth Attack: *Linking Instabilities*

Preface

Eunhoo Lee

Just as the approach to the land of Yongsan started and ended with water, I invited people to understand the work within the sequence. I would help them gradually face the fact that there is a line on the time axis and a considerable architectural challenge waiting for them on the other side. This is also how I embraced the discipline of architecture during my five year studies. There's the freshman, when you enter with a vague mind and pure thoughts, and then there's the junior, when you're physically struggling and relishing the pain and joy at the same time. In an instant, graduation unveils. It's a very momentary and intense time. I don't feel like it's over, but still living it. I think we all would not be able to get out of these days easily.

- Because it was intensely hard but fun (or so I still try to think)
- Because it's a continuation of learning after all
- Because my fifth grade work files are still scattering on my desktop.

As I recall, I feel the strong thrill of landing once again coursing through my body. Waiting to reunite the scattered pieces of our rocket somewhere in the world has become a true bliss.

한강에서부터 용산 땅까지의 접근이 물로부터 시작해 물로 끝났던 것처럼, 사람들도 이 시퀀스 안에서 작품을 이해하길 바랐다. 시간 축 위에 선이 있고 그 너머에서 그들을 기다리는 상당히 건축적인 도전을 서서히 마주하도록 돕고자 했다. 학교를 다니는 5년간 건축이라는 학문을 받아들인 방법이기도 하다. 모호한 마음으로 발을 들이고 순수하게 사유하던 저학년이 있었고, 제법 물리적인 고민을 하며 고통과 재미를 동시에 향유하다가, 눈을 뜨니 졸업이다.

아주 순간적이며 강렬한 5년이다. 마무리되었다는 느낌보다는 여전히 그 시절을 살고 있는 기분인데 아마도 우리 모두 이 시간을 쉽사리 빠져나올 수 없겠다는 생각을 한다. 그 이유는,

- 격렬히 고됐지만 즐거웠기 때문 (혹은 그렇게 생각하려 노력했던 습성이 아직도 남아서)
- 결국 그 이후도 배움의 연장이기에
- 5학년 작업파일들이 아직도 바탕화면에 널브러져 있다

이렇게 지난 시간을 떠올려 문장을 되뇌니 다시 한 번 착륙의 강한 전율이 온 몸을 타고 흐른다. 여기저기 흩어진 기체 조각을 세상 어딘가에서 재회할 때를 기다리는 일이 진정 기쁨이 되었다.

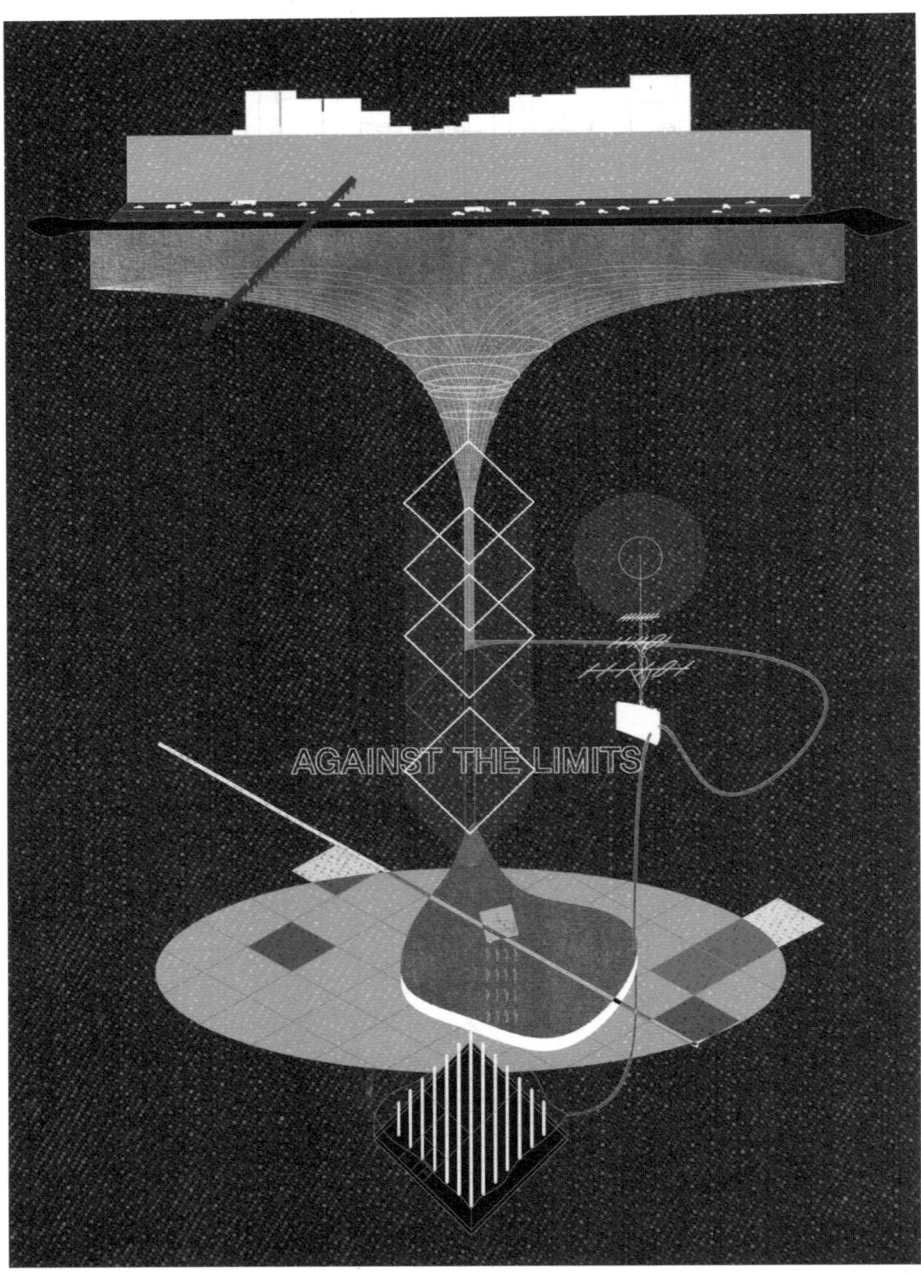

In 1887, Chief Seattle wrote of his loss in a letter to the white men who occupied his village. He thought the land would stay the same, but it never did. He left his longtime home saying that there is no death, only infinite change.

Today, I pose questions to our cities. In the life cycle of architecture, structures last 300 years, objects change in an hour, and all of our assumptions are based on the belief that the land is eternal. The question of how long we can rely on this belief is slowly materializing in cities everywhere.

In August 2020, the Jamsu Bridge was closed for 240 consecutive hours. Following the closure, sections of Olympic Boulevard were also closed to traffic. With the water level of the Hangang River reaching 8.7 meters according to the watermark, many things were unusable for a while. The threat of water is gradually expanding in frequency and scope, and not only roads and parks along the Hangang River, but also urban spaces in low-lying areas are easily pushed to their limits.

Another crisis in the city comes from development. The land of Yongsan, which is a mixture of multiple identities from the past, is still the site of intersecting interests centered around a maintenance depot.

Land is not eternal. Change is constant, and despite this, people have a desire to live at their own pace. The space that cities need nowdays must be proposed in a form that makes threats and forces visible and constantly reacts to permissions and limits. This is a project that considers both underground and above ground, ways to empower the land that is waiting for another blow, and prays for the daily lives of all city members.

1887년, 시애틀 추장은 마을을 점령한 백인들에게 보내는 편지에 상실감을 적었다. 땅은 그대로일 줄 알았지만 결코 그렇지 않았다. 죽음 또한 없으며, 무한한 변화만이 계속될 뿐이라는 말을 끝으로 오랜 터전을 떠났다.

오늘날 우리 도시에 질문을 던진다. 땅은 영원할까? 건축의 생애주기에서 구조는 300년 지속되고, 물건은 한 시간 만에 바뀌기도 하는데, 모든 가정은 땅이 영원하다는 믿음에서부터 비롯된다. 이 믿음에 언제까지 의지할 수 있는지에 대한 의문은 도시라면 어디서든 서서히 구체화되고 있다.

2020년 8월, 잠수교는 240시간 연속 통제되었다. 잠수교가 통제된 뒤, 올림픽대로의 구간들도 뒤따라 통행이 제한됐다. 기준표에 따른 한강 수위가 8.7m에 육박해 한동안 많은 것을 사용할 수 없었다. 물의 위협은 빈도와 범위가 점진적으로 확장되고 있으며, 한강변의 도로나 공원뿐만 아니라 저지대의 도시 공간도 쉽게 한계에 처한다.

도시의 또 다른 위기는 개발에서 비롯된다. 과거로부터 여러 정체성이 혼합되어 나타난 용산이라는 땅은 지금까지도 정비창을 중심으로 여러 이해관계가 교차하는 모습이다.

땅은 영원하지 않다. 변화는 끝없고, 그럼에도 불구하고 사람들은 제 속도를 유지하며 생활하려는 욕구가 있다. 오늘날 도시에 필요한 공간은, 위협과 힘이 가시화 되어 허용과 한계에 지속적으로 반응하는 형태로 제안되어야 한다. 또 다른 타격을 기다리고 있는 땅에게 힘을 실어줄 수 있는 대비책을 지하와 지상에서 두루 고민하고, 도시 구성원 모두의 일상 영위를 기도하는 프로젝트이다.

1. Site as a Land
2. Site as a City
3. Intersecting Boundaries
4. Site Analysis
5. Strategy
6. Proposal
7. Scene
8. Design

Here, land is divided into two categories. First, the land as land, and second, the land as a city. In both cases, I follow the cracks of the "eternity of the land."

I chose Yongsan as the site where the aforementioned phenomenon is most evident. The goal was to analyze the different boundaries that can be found on the site, and to connect and visualize the boundary elements. The first step of visualization is to reorganize the levels of the existing city. There, thresholds are created, which serve to connect different levels, blocks of different characteristics, threats and responses, tolerance and limits.

The project begins by bringing water into the city. It utilizes a rainwater pumping station and continuous rain gutters built in 2018 to address the frequent flooding of Yongsan. It slows the water down from its fast pace through the city and collects it back into the center of the city. I consider how to utilize the existing urban infrastructure that is not operational for a few days a year and propose a space that works as a new circulation system.

여기서 땅은 두 가지로 나뉜다. 첫째는 토지로서의 땅, 둘째는 도시로서의 땅이다. 두 가지 경우에서 '땅의 영원성'에 균열이 가는 모습을 쫓았다.

앞서 말한 현상이 가장 잘 드러나는 용산을 사이트로 선정했다. 사이트에서 찾을 수 있는 여러 경계를 분석하고, 그 경계 요소들을 연결하고 가시화하는 것을 목표를 가졌다. 가시화의 첫 단계는 현존하는 도시의 레벨을 재조직하는 것이다. 그곳에는 문지방이 생겨나는데, 각각의 레벨, 서로 다른 특성의 블럭, 위협과 대응, 허용과 한계를 연결하는 역할을 한다.

프로젝트는 물을 도시 안쪽으로 끌어들이면서 시작된다. 2018년 용산의 잦은 침수를 해결하기 위해 지어진 빗물펌프장과 연속 빗물받이를 활용한다. 빠르게 도시를 통과하던 물의 속도를 늦추고, 다시 물을 도시 한 가운데로 모은다. 1년에 며칠 가동되지 않는 기존 도시 하부구조의 활용방안을 고민하고 새로운 순환체계로 작동하는 공간을 제안하고자 한다.

The Fourth Attack: *Linking Instabilities*

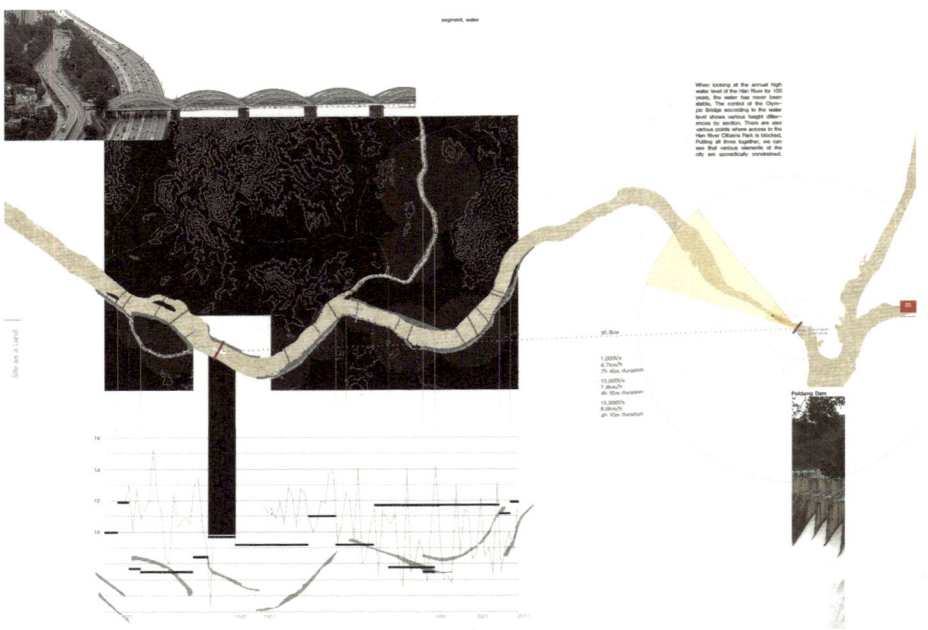

The lower part of the Hangang Bridge is deep and has been the standard for water level observation since the Japanese occupation. When the water level of the Hangang River exceeds 3.6 meters, walking on the submerged bridge is restricted, and when it exceeds 5.2 meters, the Banpo Bridge is closed. At 8.4 meters, traffic is restricted in both directions on the Olympic Boulevard. When we hear about flood warnings and flood advisories in Seoul on the news, they are all based on water level observation stations at the Hangang Bridge. The bridge has been recording the water level of the Hangang River for a long time, providing a measure of not only the height of the water but also the regulation of the city. The standard of observation has not changed over the years to ensure the continuity of the records based on this standard.

한강대교 하단은 수심이 깊어 일제강점기부터 수위 관측의 기준이 되었다. 한강대교 기준 한강의 수위가 3.6m를 넘어가면 잠수교 보행이 제한되고, 5.2m를 넘으면 반포 한강대교를 폐쇄한다. 8.4m가 넘으면 올림픽대로 한강철교 구간 양방향 차량통행이 제한된다. 우리가 뉴스를 통해 서울의 홍수주의보, 홍수경보를 통보 받을 때도 모두 한강대교의 수위관측소가 기준이 된다. 오랜 시간 한강의 수위를 기록해온 한강대교는, 물의 높이 뿐만 아니라 도시를 조절하는 척도를 제시하고 있다. 관측의 기준을 여태 바꾸지 않은 것도 이 기준을 바탕으로 하는 기록의 연속성을 위함이다.

The Fourth Attack: *Linking Instabilities*

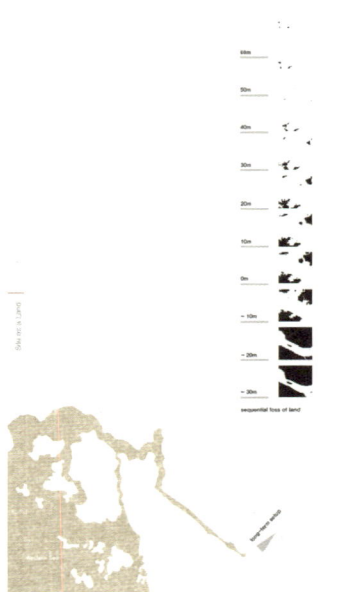

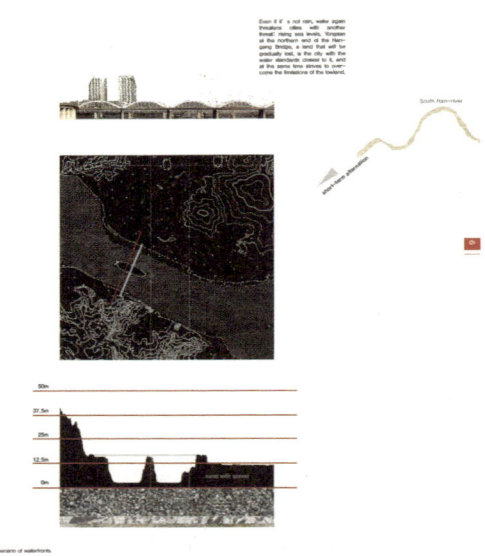

When summarizing the highest annual water level of the Hangang River for 100 years, the water has never been stable. The control of the Olympic Boulevard by the water level varies from section to section. The point at which the Hangang River Citizens' Park is closed to the public also varies widely. Putting all three together, we see that many elements of the city are irregularly and sporadically pushed to their limits.

Even if it's not rain, water threatens the city once again with the threat of sea level rise. Slowly losing land, Yongsan, at the northern end of the Hangang Bridge, is the city closest to the water's edge and struggles to overcome the limitations of low-lying land.

Seoul is unstable. During the rainy season, roads and home flood, and schools close. To be safe from the threat of water, people build pump stations near the Hangang River to spit out water that constantly seeps into the city to protect the land from water. Despite this, the city continues to flood, leaving areas unprotected.

100년 동안의 연별 한강 최고 수위를 정리해보았을 때, 물은 결코 안정적이지 않은 모습이다. 수위에 따른 올림픽대로의 통제는 구간별로 다양한 높이 차를 보인다. 한강시민공원의 이용이 차단되는 지점도 아주 다양하다. 이 세 가지를 모두 모아 보면 우리는 도시의 여러 요소들이 불규칙적이고 산발적으로 한계에 처한다는 것을 알 수 있다.

비가 아니더라도, 해수면 상승이라는 위협으로 물은 또 한 번 도시를 위협한다. 서서히 잃어갈 땅, 한강대교 북단의 용산은 물의 기준을 가장 가까이에 둔 도시인 동시에 저지대의 한계를 극복하려 애쓴다.

서울의 땅은 불안정하다. 장마철이 되면 도로가 잠기고, 집에 물이 들고, 학교가 문을 닫는다. 물의 위협으로부터 안전하기 위해 사람들은 한강 가까이 펌프장을 짓고, 땅을 물로부터 보호하기 위해 계속해서 도시 안에 스며든 물을 뱉어낸다. 그럼에도 불구하고 도시는 계속 물에 잠기고, 미처 보호하지 못하는 구역들이 생긴다.

Yongsan was the first city to be developed as the city expanded, the gateway to Gangnam, the port, and a lantern. Depending on its location and geographical characteristics, Yongsan is a mix of all these identities.

The KTX and Gyeongu Central Line railroad tracks that pass through Yongsan Station cut off the land. The Yongsan District Office is planning to create a green axis by undergrounding the overhead railroad tracks, and to integrate neighboring land to create separate complexes. Tired of development plans that are constantly being debated amidst political and social issues, I ask Yongsan the same question;

Is land eternal? There is a concern that efforts to ameliorate disconnection and discontinuity will eventually re-define the boundaries of blocks.

The Eastern Expressway cuts quickly through the city while preventing the Hangang River from flooding into the city. Apartments lining the riverfront, railroad tracks running between blocks, topographical differences disrupting circulation, and a gridlocked city that has been marginalized by all of this, keeping the innermost space from 100 years ago. It's a mixed bag.

The project flows in the direction of connecting the elements that threaten the unstable land of Yongsan, from water and development. I imagine that crisis and response are reflected in the city, which becomes a way to resolve the anxiety. It is as if human awareness leads to preparation. I envision a moment when architecture will take on the role.

용산은 도성이 확장하며 개발된 첫 번째 도시이기도 했고, 강남으로 가는 관문이었으며, 한강에 배가 들고 날 나루터였고,등불의 역할이기도 했다. 위치적, 지리적 특성에 따라 쓰임을 달리하는데, 오늘날 용산은 이렇게 과거로부터 쌓아온 정체성이 모두 혼합되어 나타난다.

용산역을 지나는 KTX와 경의 중앙선 철로로 인해 단절된 땅들이 존재한다. 용산구청은 지상 철로를 지하화하여 녹지 축을 만들고, 인근 토지를 통합하여 각각의 복합 단지를 조성하려는 계획 아래 있다. 정치적, 사회적 이슈 속에 끊임없이 대두되는 개발계획에 지친 용산 땅에 같은 질문을 던진다.

땅은 영원한가? 단절과 불연속성을 개선하려는 노력들이 결국 다시 블럭의 경계를 분명히 한다는 우려가 뒤따른다.

동부간선도로는 한강이 도시로 범람하는 것을 막는 동시에 빠르게 도시를 통과한다. 강변을 따라 늘어선 아파트, 블럭 사이를 가로지르는 철로, 지형 차로 인한 동선의 파괴, 그리고 이 모든 것들로 인해 소외되어, 100년 전부터 가장 안쪽 자리를 지키는 격자형 도시. 혼재의 양상이다.

본 프로젝트는 물과 개발로부터 불안정한 용산이라는 땅에 위협이 되는 요소를 하나로 잇는 방향으로 흐른다. 위기와 대응이 도시에 그대로 드러나서, 땅이 영원하지 않을 것이라는 불안이 해소될 방법이 되는 상상을 한다. 인간의 인지가 준비로 이어지는 것과 같다. 그 역할을 건축이 감당하게 될 순간을 그려본다.

The Fourth Attack: *Linking Instabilities*

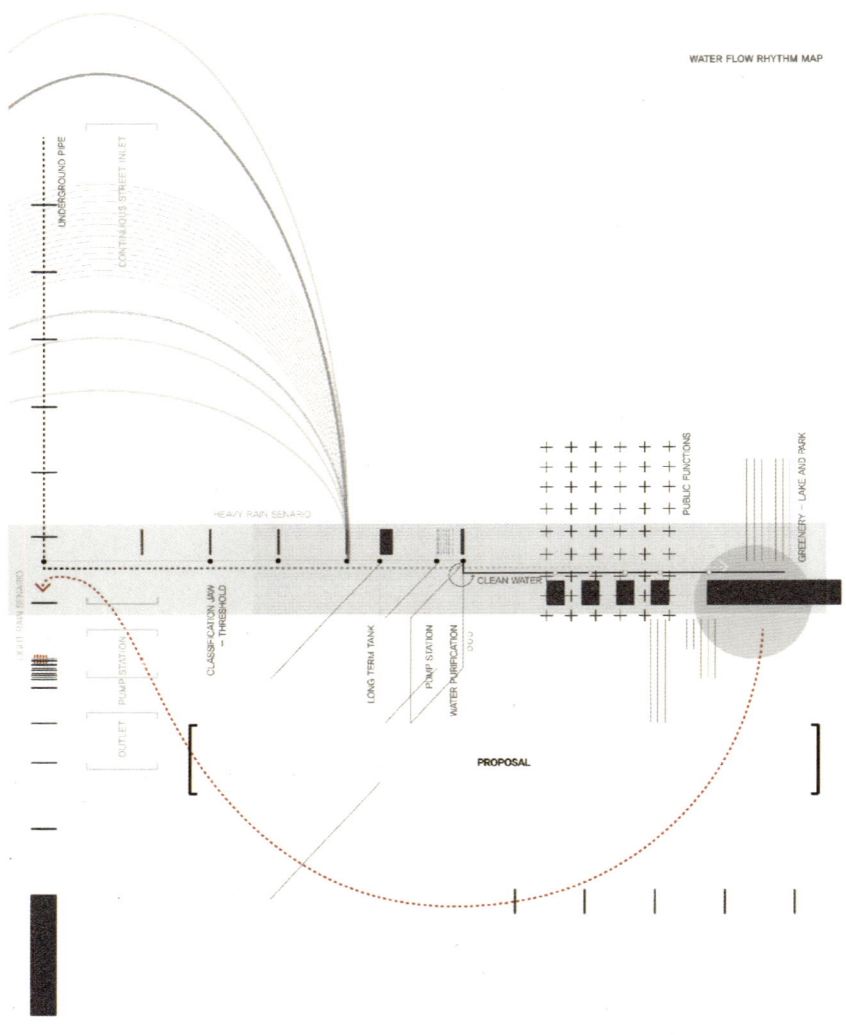

Mass Study — Eunhoo Lee

Create a new terrain inside the flat land at level 0. The terrain will act like a well, collecting the rainwater we've been so busy avoiding. By creating two concave features that collect water, and convex hills around them, it creates a more natural break in the existing surrounding terrain. The collected water flows back and forth between the contour lines, erasing the boundaries of the site and responding to the weather as it fills and empties.

0레벨의 평평한 대지 안쪽에 새로운 지형을 만든다. 지형은 곧 우물처럼 작용해 우리가 여태 피하기 바빴던 빗물을 모은다. 물이 모이는 오목한 지형을 두 개 만들고, 그 주위로 볼록한 언덕이 생겨나면서 기존 주변지형이 갖던 단차가 한층 자연스러운 형태로 나타난다. 모인 물은 콘타라인 사이를 오가며 사이트의 경계를 지우고, 날씨에 따라 차고 비워지면서 변화에 대응한다.

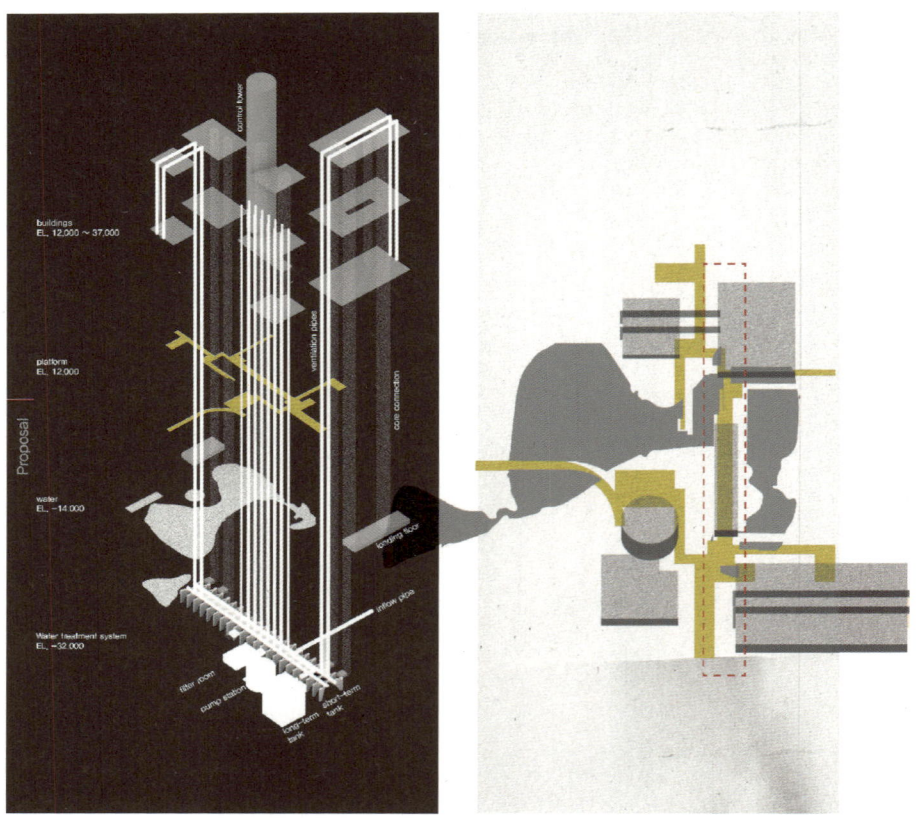

A reservoir is inserted underground to slow down rainwater. The large reservoir, 36 meters wide, 200 meters long, and 20 meters high, covers an area from the top of the existing delta to the southern end of Yongsan. The reservoir fills up during big rains and empties during the dry season, drawing programs above ground inside. Unlike the rainwater pump station, which only operates eight days a year, the reservoir is used both when it rains and when it doesn't, helping to overcome the limitations of space.

Rainwater is directed to the Hangang River via a continuous rain gutter, which was created with the addition of the rain pump station in 2018, before entering the site's basement via the sewer jaw. Rainwater flows into a long-term storage tank, and in the event of a large rainfall, a short-term storage tank is opened to store water. The collected rainwater is pumped up to ground level through a pumping room and discharged into the site's artificial lake through a filter room and discharge tank.

빗물의 속도를 늦출 용도의 저수조를 지하에 삽입한다. 폭 36미터, 길이 200미터, 높이 20미터의 큰 저수조는 기존 삼각지 상부부터 용산 남단을 모두 수혜하는 면적이다. 저수조는 큰 비가 올 때 가득 찼다가, 비가 오지않는 건기에는 비워져 땅 위의 프로그램을 안쪽으로 끌어들인다. 일년에 8일만 가동되는 빗물펌프장과 달리 비가 올 때, 오지 않을 때 모두 사용되어 공간의 한계를 극복하도록 돕는다.

빗물은 2018년 빗물펌프장을 신설하며 함께 형성한 연속 빗물받이를 타고 한강으로 향하다가, 관거 턱을 타고 사이트 지하로 유입된다. 빗물은 장기저수조로 흐르고, 큰 비가 올 경우에는 단기저수조를 열어 물을 보관한다. 모인 빗물은 펌프실을 통해 지면 레벨로 올라가, 필터실과 토출수조를 거쳐 사이트의 인공호수로 배출된다.

The Fourth Attack: *Linking Instabilities*

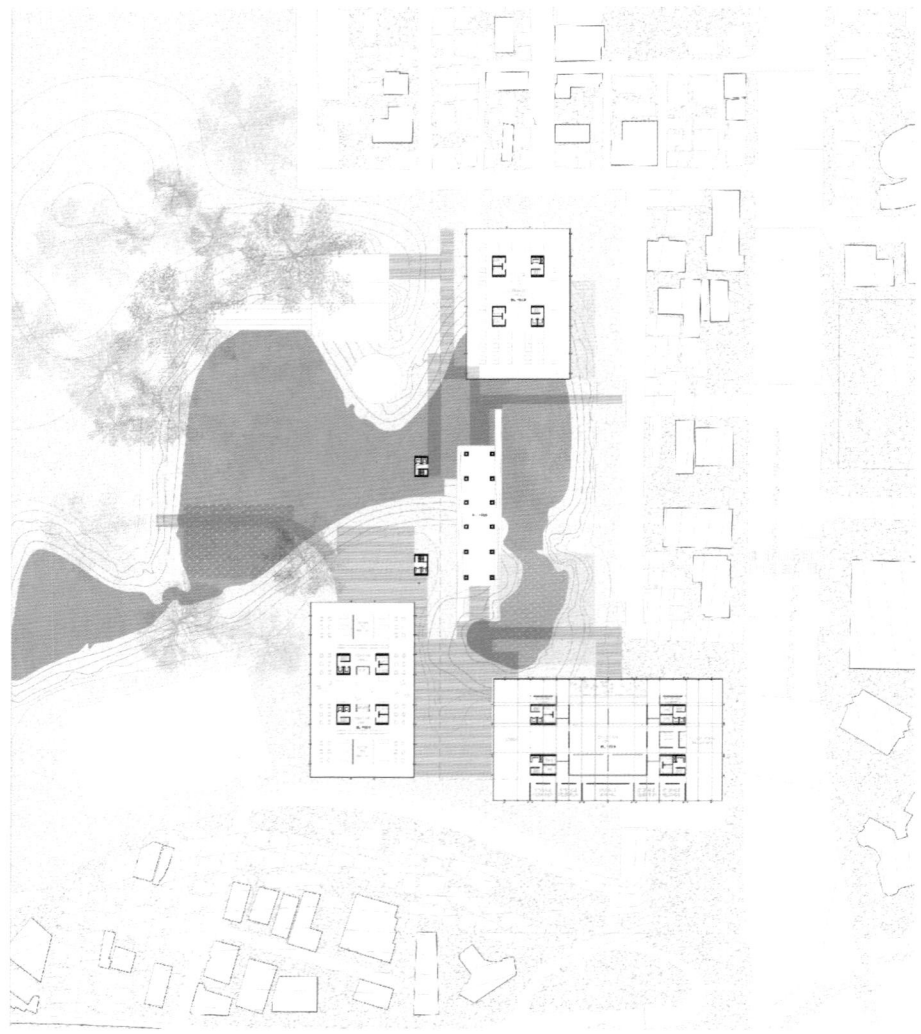

Site Plan — Eunhoo Lee

Above the lake, mass is placed to hold the programs, with platforms above the water connecting them. The above-ground programs of the pool, market, and gallery are connected by platforms, which in turn extend underground to reservoir during the dry season. Freed from its limits, the site absorbs its surroundings inward. A terrain is created to accommodate the water, and a program is built around the terrain, connecting to the road to the west and extending along the platform. It is a way to provide a new form of daily life for the small lattice dwellings of the fading Yongsan.

호수 위로는 프로그램을 담을 매스를 배치하고, 이를 연결하는 물 위의 플랫폼을 얹는다. 수영장, 마켓, 갤러리의 지상 프로그램은 플랫폼으로 연결되고, 또 지하로 확장되어 건기에는 저수조까지 연장되는 형태다. 한계로부터 자유로워진 사이트는 주변 상황을 안쪽으로 흡수한다. 물을 수용하는 지형이 생기고, 지형 주위로 프로그램이 생기고, 서쪽의 도로와 연결되어 플랫폼을 따라 뻗어나간다. 얼마 남지 않은 용산의 작은 격자형 주거지에 새로운 형태의 일상을 제공하는 방법이다.

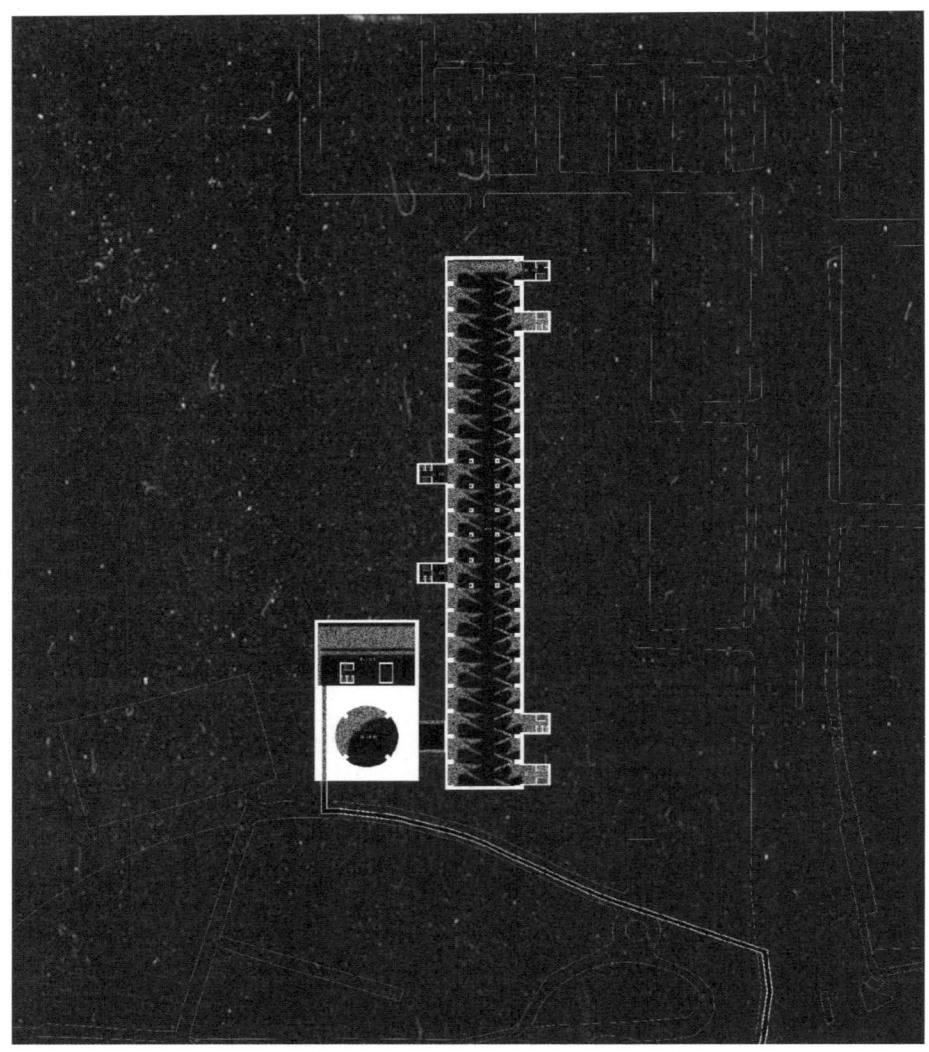

On a rainy day, the water in the tank flows back to the Hangang River through the pipe.

비가 많이 오는 날에는 저수조를 채운 물이 다시 관거를 타고 한강으로 흐른다.

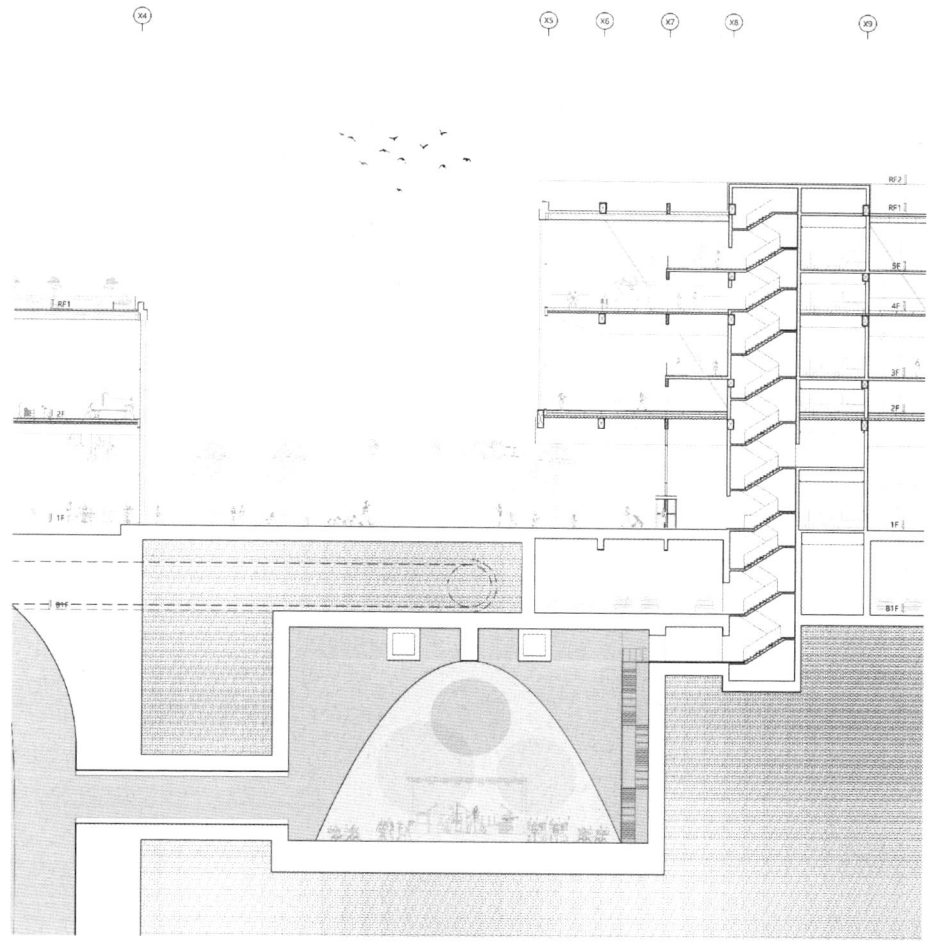

On dry season, the program on the ground flows into the short-term tank. Imagine residents enjoying performances, markets, and sports together in gigantic underground space.

비가 오지 않을 땐 지상의 프로그램이
단기저수조로 흘러 들어간다.
지하의 대공간에서 주민들이 함께 공연,
마켓, 스포츠를 즐기는 모습을 상상한다.

Perspective Drawing   Eunhoo Lee

The Fourth Attack: *Linking Instabilities*

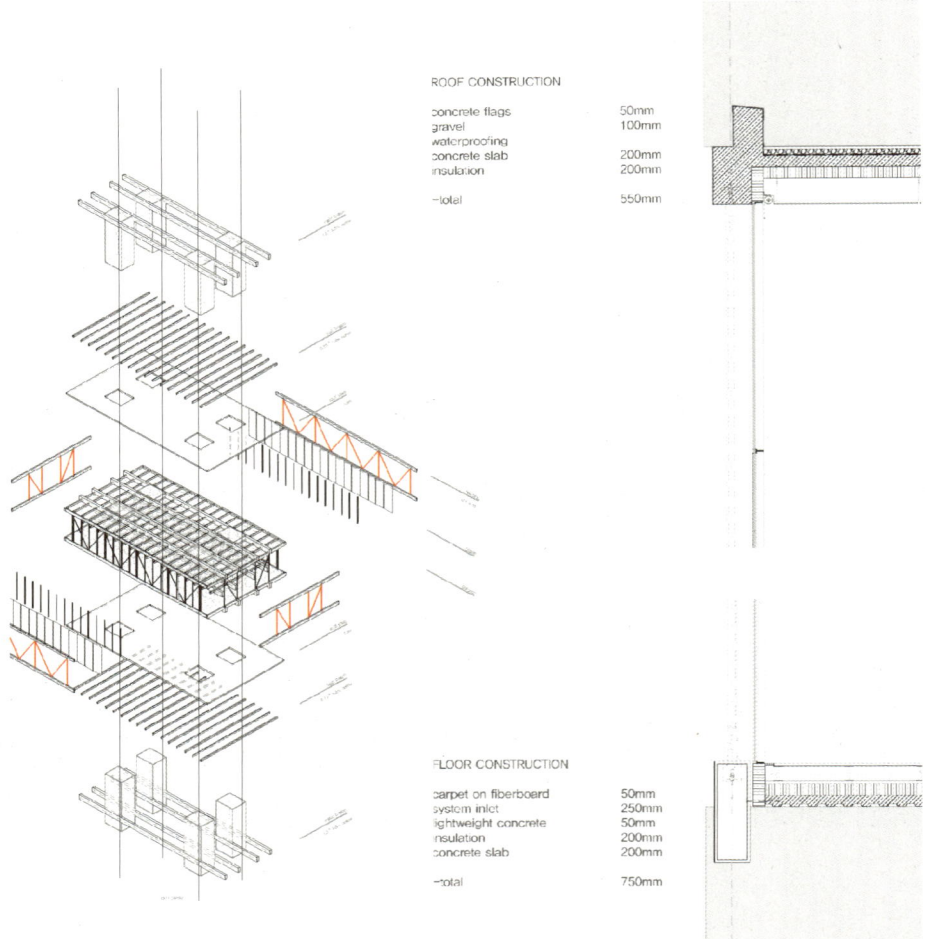

ROOF CONSTRUCTION

concrete flags	50mm
gravel	100mm
waterproofing	
concrete slab	200mm
insulation	200mm
–total	550mm

FLOOR CONSTRUCTION

carpet on fiberboard	50mm
system inlet	250mm
lightweight concrete	50mm
insulation	200mm
concrete slab	200mm
–total	750mm

The structural system of the main mass lifts the building off the ground and allows for flexible flow within the site. The rectangular exhibition hall mass floats 6 meters above the ground. Two large hollow beams measuring 0.6m by 1.8m make up the structure, each with two main beams forming a □-frame with cables to support the roof and lower slabs of the pavilion.

The four cores share the structural frame and the 63-meter long span, while also serving as the main circulation and utilities that connect the underground cistern to the exterior and up to the upper mass at once. This gives the building a free plan with a short span of 10 meters and a long span of 28 meters, which reduces interference from the ground and suggests that the upper exhibition program can experiment within the plan.

Along the building's envelope, columns spaced 3.6 meters apart compensate for the bending moments in the beams and the slab's outer structure, and outside the columns, cables are coiled every 7.2 meters to tie the building back together. This allows the floating mass measuring 34m by 63m to function transparently and lightly. Here, the 3.6-meter column spacing and 7.2-meter wire spacing act as modular units inside and outside the building, giving the large mass a series of rhythms.

The placement of the exhibition hall, at the innermost part of the plan, opens the building's exterior to the outside, adding to the elevational autonomy of the structural system. A glass curtain wall is applied to the roof and lower slab, which is suspended in the form of a counterweight, to increase the tightness of the insulation.

메인 매스의 구조 시스템은 건물을 지면에서 들어올려 사이트 내부로의 흐름이 유연토록 돕는다. 직사각형의 전시관 매스는 지상에서 6미터 위에 떠있다. 2개의 0.6미터 × 1.8미터의 커다란 중공 빔이 구조를 구성하는데, 각각 2개의 메인 빔이 케이블과 함께 ㅁ자 프레임을 형성하여 전시관의 지붕과 하부 슬라브를 받치고 있다.

4개의 코어는 지하 저수조로부터 외부를 거쳐 상부 매스까지를 한 번에 연결하는 주요 동선과 설비의 역할인 동시에 구조 프레임과 63미터의 긴 스팬을 분담한다. 따라서 건물의 짧은 스팬은 10미터, 긴 스팬은 28미터로 자유로운 평면을 갖추게 되어 땅으로부터의 간섭이 줄어들고, 상부의 전시 프로그램이 플랜 안에서 다양한 시도를 할 수 있도록 제안한다.

건물의 외피를 따라 3.6미터 간격의 기둥이 빔에 가해지는 휨 모멘트와 슬라브 바깥쪽 구조를 보완하며, 기둥 밖으로는 7.2미터 간격으로 케이블을 감아 건물을 다시 하나로 엮는다. 이는 34미터 × 63미터의 떠있는 매스를 투명하면서도 가볍게 기능하도록 한다. 여기서 3.6미터의 기둥 간격과 7.2미터의 와이어 간격은 건물 안팎의 모듈 단위로 작용하며 넓고 큰 매스에 일련의 리듬감을 부여한다.

전시장의 배치는 평면 가장 안쪽에 자리해 건물의 바깥쪽이 외부로 열리게 되는 상황에 구조 시스템이 주는 입면의 자율성이 더해졌다. 역보 형태로 매달린 지붕과 하부 슬라브에 유리 커튼월을 적용하면서 단열의 기밀성을 높이고자 했다.

The Fourth Attack: *Linking Instabilities*

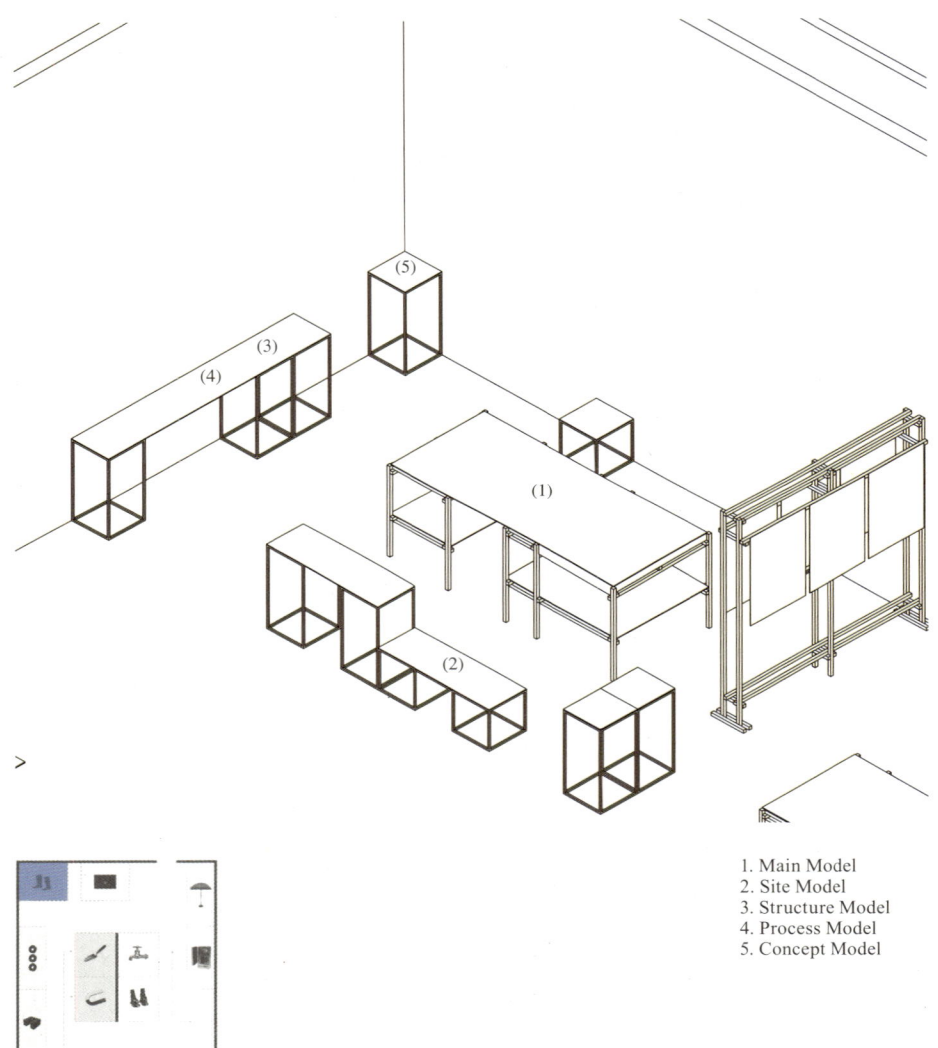

1. Main Model
2. Site Model
3. Structure Model
4. Process Model
5. Concept Model

졸업도우미

김민혁
김철휘 박유신 송윤선 표준우
박준성 최지아

MAIN MODEL

Scale 1:300
3D Print PLA + Acrylic + Wire + Paper on Woodrock, 120 by 240 cm

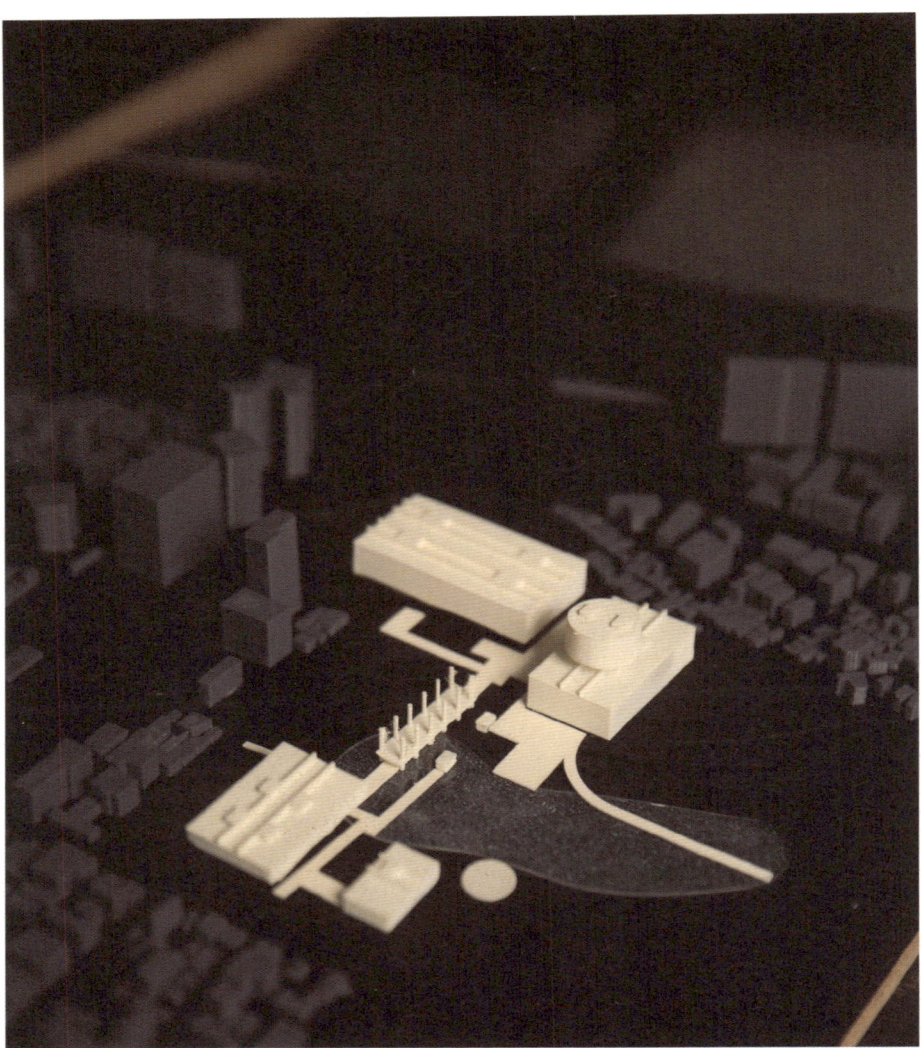

SITE MODEL

Scale 1:2000
3D print PLA + Paper + Acrylic, 40 by 120 by 10 cm

STRUCTURE MODEL

Scale 1:300
3D print PLA + Wire + Paper, 30 by 15 by 14 cm

| | Plowing and Blowing | Seungmo Seo Studio |

이지윤
Jiyun Lee

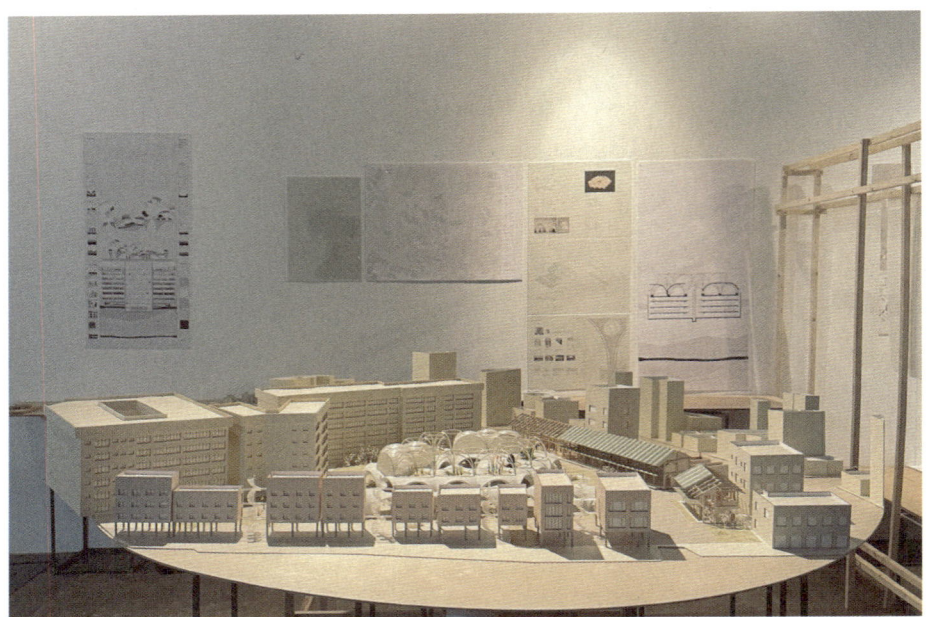

Preface

Jiyun Lee

There is architecture as we usually think of it, and there is architecture that can only exist in our imagination. Since studying architecture as a discipline, I have been struggling with the gap between my desire to design buildings that I can see and touch with my own eyes, and architectural work that cannot be built under realistic conditions but can contain my ideals and wishes. During the year-long design process at school, I went through a series of steps over the course of ten months, starting with intensive research to give my work a logic and rationale. In the process, I was troubled and painfully aware that the shape of the building I envisioned in my head at the beginning of the project and the shape that actually appeared at the end of the project were very different. I asked myself how I could preserve the original lines and feel.

The compromise I reached was to appreciate the unconstrained conditions and situations that only school can provide, and that I can fully enjoy when working as a student, and to not be overwhelmed by logic and justification, and to hope that the resulting work will be good when it is actually built.

My work had in common a different way of seeing architecture. They were all about rethinking the basic premises and definitions, and taking a fresh look at the role and definition of architecture. I questioned the relationship between the earth, architecture, and humans, from soil to massive infrastructure, and suggested attitudes that architecture could take in between.

For the last work, I wanted to finish the land that I wanted to deal with, utilizing the feeling and lines that I first imagined. Of course, as always, there were points that changed and changed, but I would definitely like to visit it if it was actually built.

우리가 보통 생각하는 지어지는 건축과, 상상 속에만 존재할 수 있는 건축이 있다. 건축을 학문으로 공부한 이래로 내 안에서는 직접 눈으로 보고 만질 수 있는 건조물의 설계에 대한 욕망과, 현실적인 여건으로 지어지지 못하지만 나의 이상과 바람을 담을 수 있는 건축작업 사이의 간극에서 고민이 많았다. 학교에서 1년간 진행되는 설계의 프로세스들은 열달의 시간만큼 켜켜이 쌓인 단계를 밟고, 밀도 있는 리서치 작업으로부터 시작해 작업에 논리와 당위성을 부여하는 과정이었다. 그 과정 속에서 처음 프로젝트를 시작할 때 내 머릿속에 그려진 건물의 형상과, 프로젝트가 마무리되었을 때 실제로 나타난 형상이 너무나 달랐기에 고민되는 부분도 많았고, 고통스럽기도 했다. 처음의 선과 느낌을 살릴 수 있는 방법이 무엇인지 질문했다.

고민 끝에 도달한 타협점은, 학교에서만 다룰 수 있는, 학생 작업일 때 충분히 누려볼 수 있는 제약 없는 조건들과 상황에 감사하자는 것이었다. 그리고 논리와 당위성에 압도되지 않고 도출된 작업의 결과물이 실제로 지어졌을 때도 좋은 작업이었으면 한다는 지점이었다.

내 작업들은 건축을 다른 방식으로 본다는 공통점이 있었다. 기본 전제와 정의에 대해 생각해보며 건축의 역할과 정의에 대해서 새롭게 바라보는 작업들이었다. 토양에서부터 거대한 인프라까지 지구와 건축, 그리고 인간 사이의 관계에 반문하며 그 사이 건축이 취할 수 있는 태도를 제안했다.

마지막 작업은 다뤄보고 싶었던 땅에, 처음 상상한 느낌과 선을 살려 마무리하고 싶었다. 물론 언제나 그랬듯이 틀어지고 달라지는 지점들은 있었지만 실제로 지어진다면 꼭 한번 가보고 싶다.

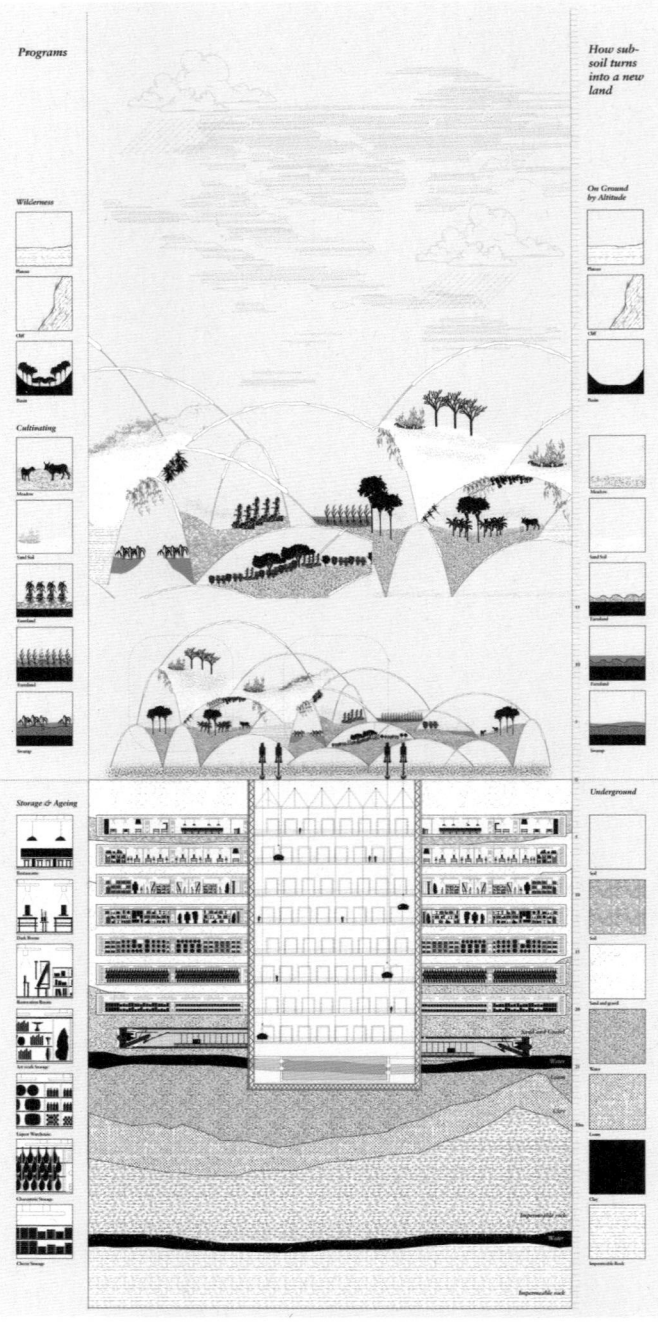

While the act of building usually means laying a foundation on the ground to provide a safe place for people to settle, this project reimagines architecture as a role that involves not only erecting a building that will not collapse on a secure foundation, but also planting pillars to enrich the lives of the creatures living on the earth. It is approached from the perspective of agriculture as an active agent that creates life. I thinned the land of Samgak-ji, where densification and fossilization are in progress, and added the concept of wind to revitalize the soil.

The ultimate goal is to revitalize the urban land that has been contaminated and hardened by unreasonable construction activities, starting from healthy soil, and adopts a strategy that integrates construction activities and results in this process.

In translating agricultural-scale technologies to the urban scale, different civil and architectural techniques were applied to the design by dividing the superstructure and substructure. In the process of excavating the underground layer, a tunnel method is used to create a large underground space, while the upper structure is stacked in the form of a dome, and the soil dug from the underground is stacked between the dome structures to create an uneven soil core to secure the diversity of vegetation.

건축행위란 일반적으로 땅 위에 기초를 세워 사람이 정주할 수 있는 안전한 실을 얻는 것을 의미하지만, 이 프로젝트는 건축을 안전한 지반 위에 무너지지 않을 건물을 세우는 것이 아닌 기둥을 박아 지구에 사는 생물의 삶을 풍요롭게 가꾸는 역할로서 새롭게 접근한다. 생명을 탄생시키는 능동적 주체로 농법의 관점에서 접근한다. 고밀화, 화석화가 진행중인 한강대로의 한 켠에 위치한 삼각지의 땅을 솎아서 바람을 넣는다.

건강한 토양으로부터 시작되는 도심 속의 흙고르기, 수없는 건축행위로 오염되고 굳어진 도시의 땅을 회복시키는 것이 궁극적 목표이며, 이 과정에 건축행위와 결과물을 통합시키는 전략을 취한다.

농업 스케일의 기술을 도시 스케일로 치환할 때, 상부 구조와 하부구조를 나누어 상이한 토목, 건축 기법을 설계에 적용했다. 지하층을 파내는 과정에서는 하부에 갱도를 먼저 조성하는 터널 공법을 통해 지하 대공간을 조성하고, 상부 구조는 돔이 적층되는 형태로 돔 구조의 사이에 지하에서 파낸 흙을 쌓아 일정하지 않은 토심을 조성하여 식생의 다양성을 확보하는 전략을 취했다.

1. Thesis
    Piles of Earth & Tree-like Architecture
    땅에 대한 새로운 접근과 건축행위에 대한 반론

2. Site Analysis in Urban Context
    Hole in City & Contaminated Soil
    도시적 접근: 오염되어 회복이 필요한 땅

3. Research Book
    Land
    땅과의 관계: 오염된 땅을 회복하는 방법들

4. Form & Structural Study
    Structure
    구조와의 관계: 나무와 같은 건축

5. Plan. Section & Perspective Drawing
    Urban Context
    경작지와의 관계: 도심에 심어진 1차 산업

6. Details & Model

Process — Jiyun Lee

Architecture is usually thought of as the act of erecting a building that will not collapse on a secure foundation. However, in this definition, the ground is considered a mere background and a passive entity, but in this project, the ground is approached from the perspective of agriculture as an active entity that gives birth to life. Architecture is approached in a new way as a role that enriches the lives of the creatures living on the earth by driving pillars into the ground.

In order to find a site where this new definition of land and architecture could be realized, I researched land that had been contaminated by urban approaches and needed to be rehabilitated. I chose Samgak-ji, a huge infrastructure facility where four railroad tracks cross the Korean Peninsula, and a land contaminated by a military facility that was stationed there for 60 years, and focused on three relationships.

The first is the relationship between architecture and the land, and I researched methods of rehabilitating contaminated land, focusing on farming methods that rehabilitate hardened land, researching examples of digging techniques and artificial terrain, and analyzing the soil of the site.

The second was the relationship between architecture and structure, thinking about how the physical methods of restoring contaminated land could be implemented structurally. As a result, I proposed a process in which the excavated soil rises to the top and is replanted into the soil of the cultivated land, and crops grow from the new soil, just as trees absorb nutrients through their roots and the nutrients travel up the trunk, causing leaves to bloom and grow.

The third was the relationship between the architecture and the cultivated land, considering the relationship between the buildings placed within the site and how a primary industry planted in the city center could function, as well as the urban context of the program and its surroundings. I wanted the low-slung primary industry to permeate the dense Hangang-daero and loosen up the rigid city center.

건축은 보통 안전한 지반 위에 무너지지 않을 건물을 세우는 행위로 생각되어진다. 그러나, 이러한 정의에서 땅은 그저 배경이자 수동적인 존재로 간주되지만, 이 프로젝트에서는 땅을 생명을 탄생시키는 능동적 주체로 농법의 관점에서 접근한다. 건축은 이러한 땅에 기둥을 박아 지구에 사는 생물의 삶을 풍요롭게 가꾸는 역할로서 새롭게 접근한다.

새롭게 정의된 땅과 건축을 실현할 수 있는 사이트를 찾기 위해 도시적인 접근으로 오염되어 회복이 필요한 땅을 리서치했다. 한반도를 가로지르는 네 개의 철로가 만나는 거대한 인프라 시설과, 60년간 주둔한 군시설로 오염된 땅인 삼각지를 사이트로 정했고, 세 가지 관계에 주목했다.

첫 번째는 건축과 땅과의 관계로, 오염된 땅을 회복하는 방법들을 리서치했다. 굳어버린 땅을 다시 회복하는 농법에 주목하며, 땅을 파는 기법과 인공 지형의 사례를 리서치하고, 대상지의 토양을 분석했다.

두 번째는 건축과 구조의 관계로, 오염된 땅을 회복하는 물리적인 방법을 구조적으로 어떻게 구현할 수 있는지에 대한 고민이었다. 그 결과, 나무가 뿌리를 통해서 영양분을 흡수하고 영양분이 줄기로 타고 올라와 잎이 피고 자라나는 것과 같이 파내어진 흙이 상부로 올라와 경작지의 땅에 다시 심어지고, 새로운 흙으로부터 농작물들이 자라나는 프로세스를 제안했다.

세 번째는 건축과 경작지와의 관계로, 도심에 심어진 1차 산업이 어떻게 작동할 수 있는지 사이트 내부에 배치된 건물들의 관계와 그 프로그램과 주변의 도시적인 맥락을 고려했다. 고밀화된 한강대로의 한켠에 낮게 자리하는 1차 산업이 스며들며 경직된 도심을 풀어줄 수 있는 작업이 되길 원했다.

260 Plowing and Blowing

풍백
농법에서 -
도시까지

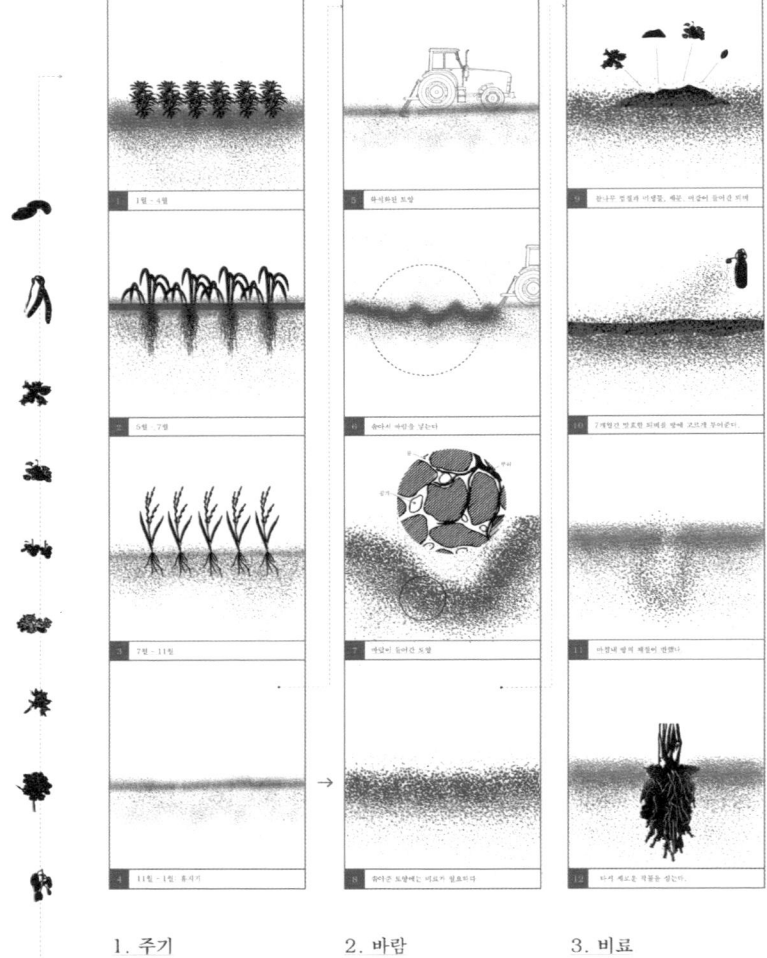

1. 주기   2. 바람   3. 비료

When it comes to farming, one of the traditional wisdoms for keeping the soil alive is the concept of Pung-Baek (風伯). The idea is that the wind is responsible for circulating the air in the soil, bringing it back to life. The geography, climate, nature of the soil, and cultivation methods used to grow the same crops can make a difference in the flavor of the produce, and applying the concept of Pung-Baek can be a way to compensate for harsh soil or climate conditions.

농사를 지을 때, 전통적으로 흙을 살리는 지혜 중에 풍백(風伯)이라는 개념이 있다. 땅속에 바람을 집어넣어 생명력을 불어넣는데, 토양 속에 공기를 순환시켜주는 역할을 바람이 담당한다는 것이다. 같은 작물을 심더라도 지리와 기후, 토양의 성질과 재배법에 따라 자라난 작물의 맛이 달라지는데 풍백 개념의 적용이 척박한 토양이나 기후조건을 보완하는 방식이 될 수 있다.

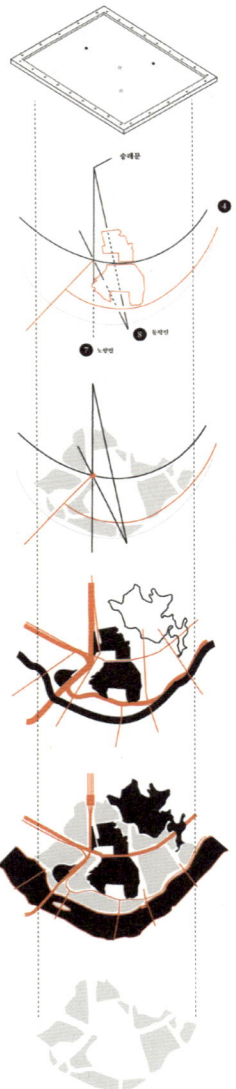

당구의 전략

당구의 전략은 상대의 궤적을 끊어내는 것이다. 상대의 수구가 적구를 향해 움직이려는 궤적을 미리 예측하여 그 동선을 끊어내는 전략으로 게임에서 승리한다. 용산의 지형적, 역사적인 궤적이 끊긴 원인도 누군가가 그 흐름을 미리 예측하여 동선을 끊어내는 전략을 취했기 때문이었다.

끊어진 옛 길

조선시대 한양에서 경상도, 충청도, 전라도로 향하던 4,7,8대로는 모두 삼각지 일대를 거쳐갔다. 옛 길을 경부선과 경의선, 미군기지와 군사시설이 끊게 되며 용산의 각 동네들은 섬처럼 산산조각 났다. 신용산와 삼각지 일대에는 아직도 옛 길의 흔적이 남아있다.

삼각지

삼각지 일대는 궤적이 끊긴 삼각형의 이질적인 필지 형태가 만들어진 지역이다. 옛 한양의 대로를 근대 도로 교통체계와 군사시설이 끊던 시기에 궤적이 끊기 다시 시작되는 곳으로 1970년대까지 원형의 로터리가 존재했다. 끊긴 궤적을 단순화 시켜 위치와 원안을 대비한다.

도로 교통체계와 군사시설

시대의 궤적과 보행의 궤적을 끊는 요소는 크게 두 가지로 요약된다. 거대한 도로 교통체제와 군사시설이다. 역사 속 존재해왔던 길과 동네들은 일제강점기 때 개통된 철로와 신설된 군사기지로 인해 분절되고 고립된다.

섬을 만드는 요소들

용산의 동네들을 섬으로 만드는 요소는 면적인 요소로 한강과, 미군기지, 남산이 있으며 선형 요소로는 경부선과 경의선, 한강대로와 백범로가 있다.

대상지는 삼각지에 위치한 한국전력 용산창고 부지이며 고밀화된 한강대로의 한켠에 자리한 버려진 땅이다. 고층건물들이 들어선 한강대로와 철도역 사이에 현재 미군이 주둔해 있는 둔지산이 자리하고, 그 끝의 한켠에 대상지가 위치해있다. 삼각지는 역사의 혼란스러운 켜들로 인해 굴절되고 분절된 지역이다. 서로를 견제하듯 대치하는 끊어진 궤적들은 도시를 화석화시킨다. 한반도를 가로지르는 네 개의 철로가 만나는 용산역과, 60년간 주둔한 군시설로 오염된 토질은 회복에도 수십년의 시간이 든다.

Site + Program — Jiyun Lee

The site is the site of KEPCO(Korea Electric Power Corporation)'s former Yongsan warehouse located in a triangular area, and is an abandoned land located in the middle of the densely populated Hangang-daero. Between the skyscrapers of Hangang-daero and the railroad station is Dunji Mount, where the U.S. military is currently stationed, and the site is located on the edge of it. Samgak-ji is a region refracted and fragmented by the chaotic lights of history. The broken trajectories that stand in opposition to each other fossilize the city. Yongsan Station, where the four railroad tracks that crisscross the peninsula meet, and the soil contaminated by 60 years of military presence, will take decades to recover.

Site mapping

경상남도 밀양시에 있는 삼문동을 대상지로 설정하였다. 삼문동의 역사적 변천과 주변 환경을 고려하여 사이트 매핑을 진행하였다. 주변을 분석하고 분석한 내용을 종합하여 사이트 매핑을 진행하였다.

By comparing and analyzing the old map and the current map, the areas that can fill islands among the areas of Yongsan which are cut off by large transportation systems such without facilities, were analyzed and categorized.

1. Landscape

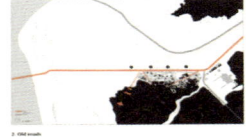
2. Old roads

3. Building types

4. Massive Transportation

5. Age of the Buildings

6. Pedestrian analysis

고지도 분석

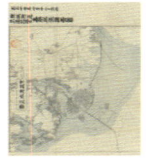
경남 수영도, 1911

경성시가전도, 1912

경성부관내도, 1917

경성도, 1913

용산시가도, 1927

Site + Program — Jiyun Lee

In order to revitalize the contaminated soil, I propose to dig a large hole in the city center to open up the strata and create a new ecosystem by thinning the soil. I did not take a historical or semantic approach to the site of the existing historical buildings and the land that needed to be restored, but applied the fundamental and simple methods of the past. I researched examples of digging techniques and artificial terrain, and analyzed the soil of the target site.

오염된 토양을 회복하기 위해 도심 속에 큰 구멍을 파내 지층을 열고 흙을 솎아 새로운 생태계를 만드는 방법을 제안한다. 기존 역사적 건물들이 있었던 곳이자 회복이 필요한 땅을 역사적이나 의미론적으로 접근하지 않고 그 이전의 근원적이고 단순한 방법을 적용했다. 땅을 파는 기법과 인공 지형의 사례를 리서치하고, 대상지의 토양을 분석했다.

고원과 분지
구름과 계곡
목초지와 백사지

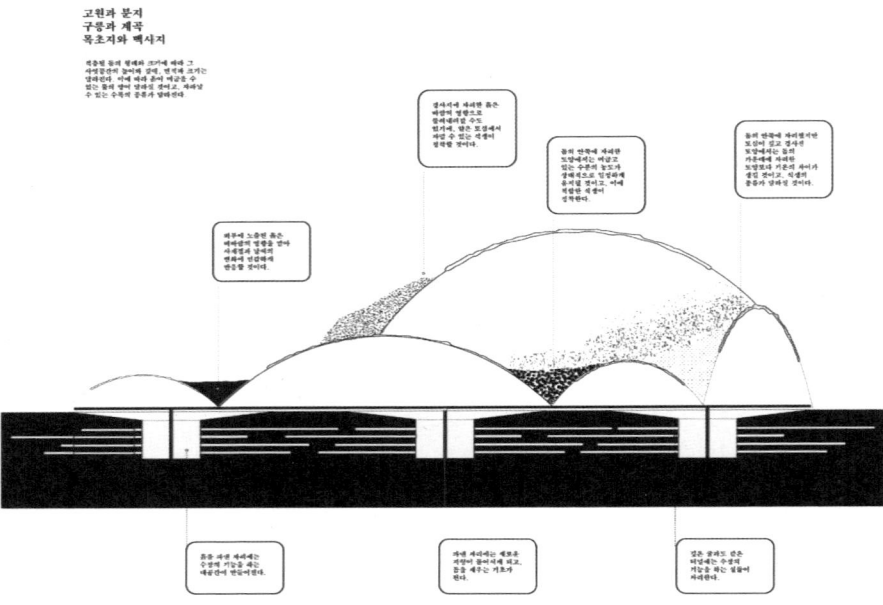

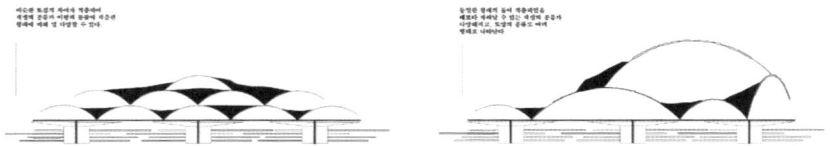

The depth of the soil core determines the types of trees that can grow. The center of a dome-shaped site will have a deep soil core. Trees with deep roots will grow from the center outward in order. Depending on the shape and size of the stacked domes, the height, depth, area, and size of the cavity will vary. This will affect the amount of water the soil can hold and the types of trees that can grow.

The excavated area is the new terrain and the foundation for the dome on top. The excavated area creates a large space that serves as a storage room.

토심의 깊이에 따라 자라날 수 있는 수목의 종류가 달라진다. 돔을 겹쳐 나온 사이공간의 한 가운데에는 토심이 깊을 것이다. 뿌리가 깊이 자라날 수 있는 수목이 순서대로 가운데에서 바깥쪽으로 자라난다. 적층될 돔의 형태와 크기에 따라 그 사이공간의 높이와 깊이, 면적과 크기는 달라진다. 이에 따라 흙이 머금을 수 있는 물의 양이 달라질 것이고, 자라날 수 있는 수목의 종류가 달라진다.

파낸 자리에는 새로운 지형이 들어서게 되고, 상층부에 돔을 세우는 기초가 된다. 흙을 파낸 자리에는 수장의 기능을 하는 대공간이 만들어진다.

1:200 structure model

Ø 100mm steel pipe

Sail vault / Pendentive

Steel pipe pendentive structure works as a roof of farmland's second floor

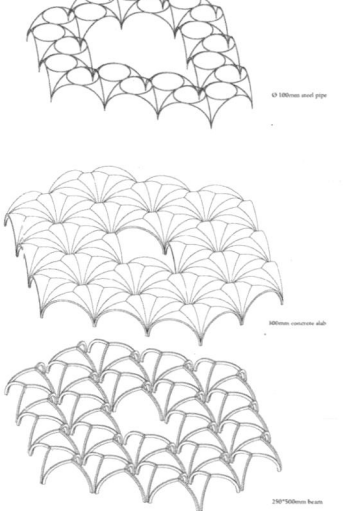

100mm concrete slab

250*500mm beam

Cross vault

250*500mm beam + 300mm concrete cross vault

Main structure is a module of cross vault, separated into two part - one for beam and another for concrete slab.

Mass Study — Jiyun Lee

Using domes and bolts as modules of the structure, I propose a tree-like architectural form. Just as firmly embedded roots draw nutrients from the soil, and water and nutrients move through the column to produce leaves, the soil of the thinned land rises to the surface and becomes the soil of the cultivated land, and the dome and bolt structure increases the surface area of the soil like a leaf, helping to create another ecosystem in the thinned area.

돔과 볼트를 구조체의 모듈로 사용하여 나무와 같은 건축적 형태를 제안한다. 단단히 박힌 뿌리가 흙 속에서 양분을 얻고 기둥을 통해 물과 양분이 이동하여 잎에서 생산활동을 하듯, 솎아낸 땅의 흙은 지상으로 올라와 경작지의 흙이 되고 돔과 볼트 구조체는 나뭇잎처럼 흙의 표면적을 넓혀 솎아져 올라온 곳에서 또다른 생태계를 만들도록 도와준다.

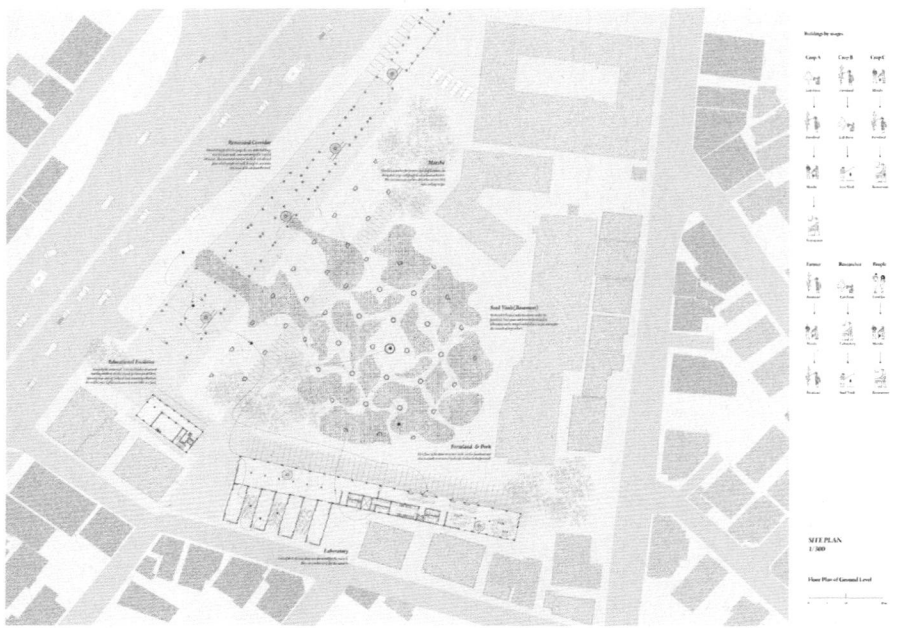

(1) Dome Structure: Cultivated Land, Farmer's Market
Dome on dome structure are used as vertical cultivated land. Depending on the diversity of the soil, different plants can be grown, and the crops grow differently depending on the season and month. Crops grown on the plots are sold, and chefs and farmers meet to share information about growing ingredients for meals.

(2) Underground: Compost Fermentation, aging, and fermentation storage
There need a place to ferment the compost we have developed and researched. Composting should be done once a year to completely change the fossilized soil. The cellars for aging and storing alcoholic beverages and dairy products are located in the basement. The warehouse for fermenting and storing fermentable crops at low temperatures is located in the basement.

(3) Renovated Building: Restaurant
A restaurant where people can enjoy dishes made from crops grown on the farm and beverages and ingredients aged in the underground aging vault. The cafeteria will be able to supply the increased demand of the business facility. It is operated as an open kitchen.

(4) Renovated Building: Educational Research Facility, Accommodation
The educational facility, where visitors can learn the techniques and stories of soil and crop cultivation, is located in a renovated building that utilizes the structure of a warehouse building. The research facility will study and experiment with seeds, seedlings, compost, and other agricultural technologies.

(1) 돔 구조: 경작지, 농부시장
돔과 돔이 얹혀진 구조를 수직 경작지로 사용한다. 토심의 다양성에 따라서 자라날 수 있는 식물이 다르고, 계절과 달에 따라 자라나는 작물이 달라진다. 경작지에서 재배된 작물들을 판매하고, 요리사와 농부의 만남을 통해서 식에 필요한 재료의 재배와 관련된 정보를 나눌 수 있는 장을 만든다.

(2) 지하: 퇴비발효장, 숙성고, 발효저장고
직접 개발하고 연구한 퇴비를 발효하는 곳이 필요하다. 화석화 된 땅의 체질을 온전히 바꾸기 위해서 1년에 한 번씩 퇴비를 만들어준다. 주류와 유제품들을 숙성하고 보관할 수 있는 창고는 지하에 자리한다. 발효하여 쓸 수 있는 작물들을 저온으로 발효하고 저장할 수 있는 창고는 지하에 자리한다.

(3) 리노베이션 건물: 식당
경작지에서 자라난 작물들과 지하 숙성고에서 숙성된 음료와 식재료들로 만든 요리를 접해볼 수 있는 식당이다. 업무시설에 따라 늘어난 식당의 수요를 이곳에서도 공급해줄 수 있다. 오픈키친으로 운영된다.

(4) 리노베이션 건물: 교육연구시설, 숙박시설
토양과 작물의 재배와 관련한 기술과 이야기들을 배울 수 있는 교육시설은 창고 건물 구조를 활용하여 리노베이션 된 건물에 위치한다. 연구시설에서는 종자와 묘목, 퇴비 등 농업에 필요한 기술들을 연구하고 실험한다.

Plowing and Blowing

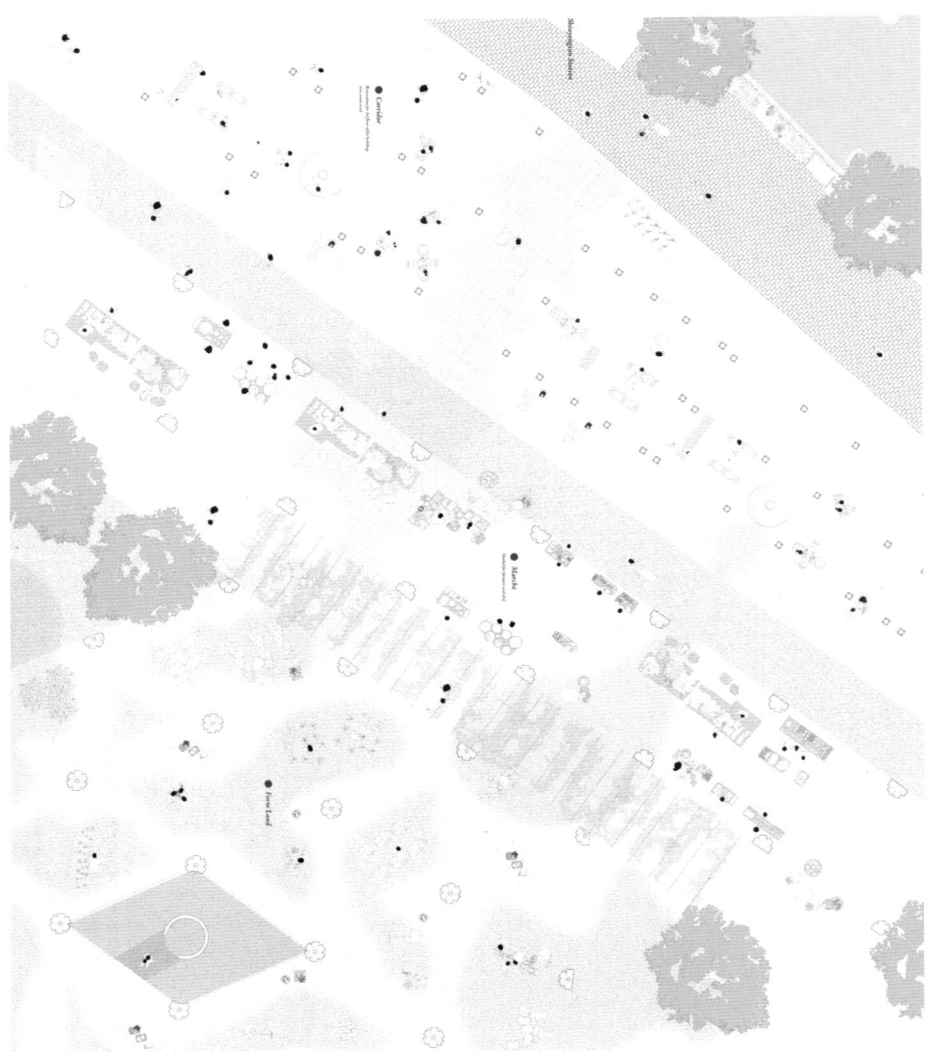

On one side of Hangang-daero, which is filled with tertiary industries and skyscrapers, there is a huge green space that is not open to the public. A military facility stands in between. A low agricultural land will be planted in Samgak-ji, where have a mix of offices, bustling restaurant alleys, military installations, massive railroad tracks and boulevards, and old housing.

3차산업과 고층건물로 가득한 한강대로의 한편에는 개방되지 않은 거대한
녹지 예정지가 있다. 그 사이를 군시설이 가로막고 있다. 오피스와
왁자지껄한 식당 골목, 군시설과 거대한 철로와 대로, 노후주거가 혼재되어
있는 삼각지에 땅과 관계하는 낮은 농경지가 스며들듯 심어진다.

Squeezed in between the inaccessible boundaries of the military installations and the vast greenery, the primary industries provide a new vitality to the city, returning to the source of life and giving pause for thought to those walking through the city.

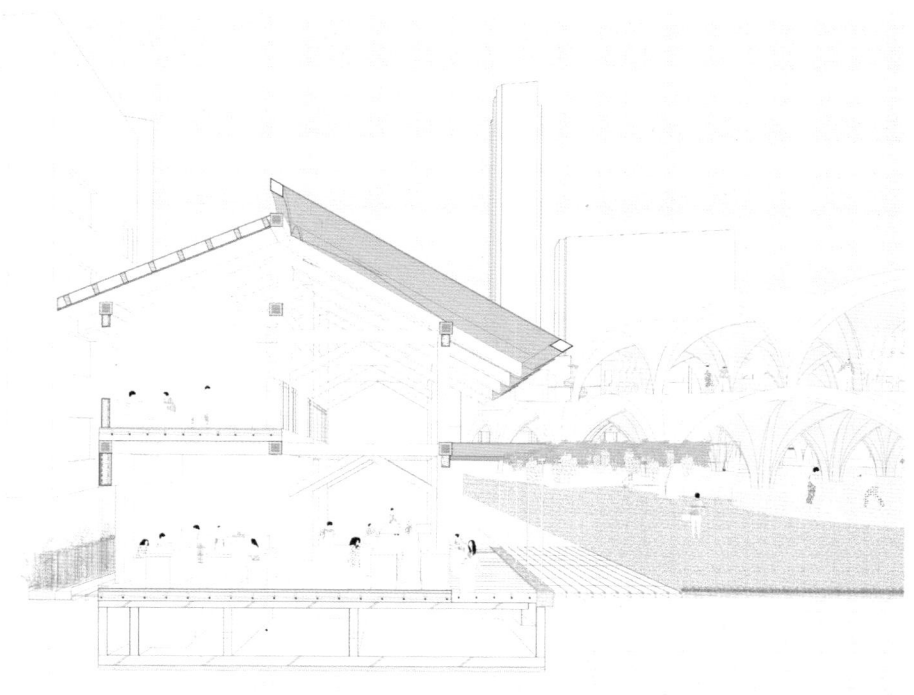

접근할 수 없는 군시설의 경계와 드넓은 녹지 사이를
비집고 들어와 자리한 1차 산업은 도시의 새로운 활력이 되어주고
생명의 근원으로 돌아가 도시를 걷는 사람들에게 생각에 잠기게 한다.

2. Construction & Details

1) Ferrocement

1. Steel supports is constructed

2. Mesh is attached to the Frame

3. Plastered on the outside with mortar

4. Surface finish

Types of construction method

No Skeletal Steel

Skeletal Steel in One Direction

Skeletal Steel in Two Directions

Combination of Mesh and Discontinuous Fibers

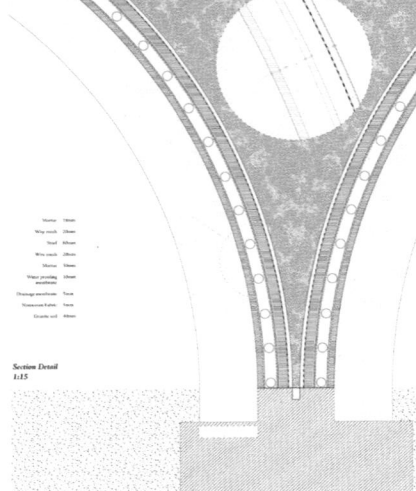

Section Detail
1:15

2) Heavy Timber structure

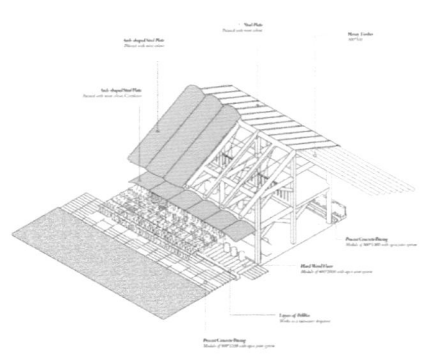

Paving Details

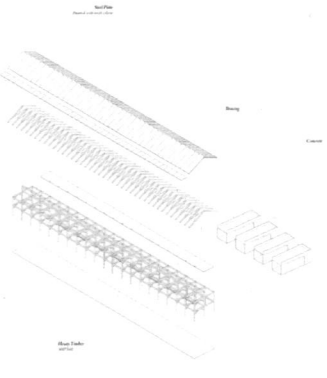

Axonometric view of structure model

Ferro-cement method was used to create the domes and vaults that would be used as agricultural fields and as modules for the structure. A high-strength cement mortar is poured over several layers of wire mesh to form the face, and on top of that a sheet for waterproofing and a layer for water supply. The innermost layer was to contain the soil, so the water circulation had to be technically addressed.

The renovated restaurant and accommodation building retains the original heavy timber structure, but adds bracing to increase structural stability and a steel plate roof so that the wooden modules are visible from the outside.

경작지로 활용될 돔과 볼트를 구조체의 모듈로 사용하기 위해 페로시멘트 공법을 적용했다. 와이어 메쉬로 여러 층을 겹친 것에 고강도 시멘트 모르타르를 부어 면을 만들고, 그 위에는 방수를 위한 시트와 급수를 위한 레이어를 더한다. 가장 안쪽에는 흙이 담겨야 하기에 물의 순환을 고려하여 기술적으로 해결해야했다.

리노베이션 되는 레스토랑과 숙박시설 건물은 중목구조의 원형을 그대로 살리되, 브레이싱을 더하여 구조적인 안정성을 높였고, 스틸 플레이트 지붕을 더해 목구조의 모듈이 외부에서도 그대로 보여질 수 있도록 했다.

Plowing and Blowing

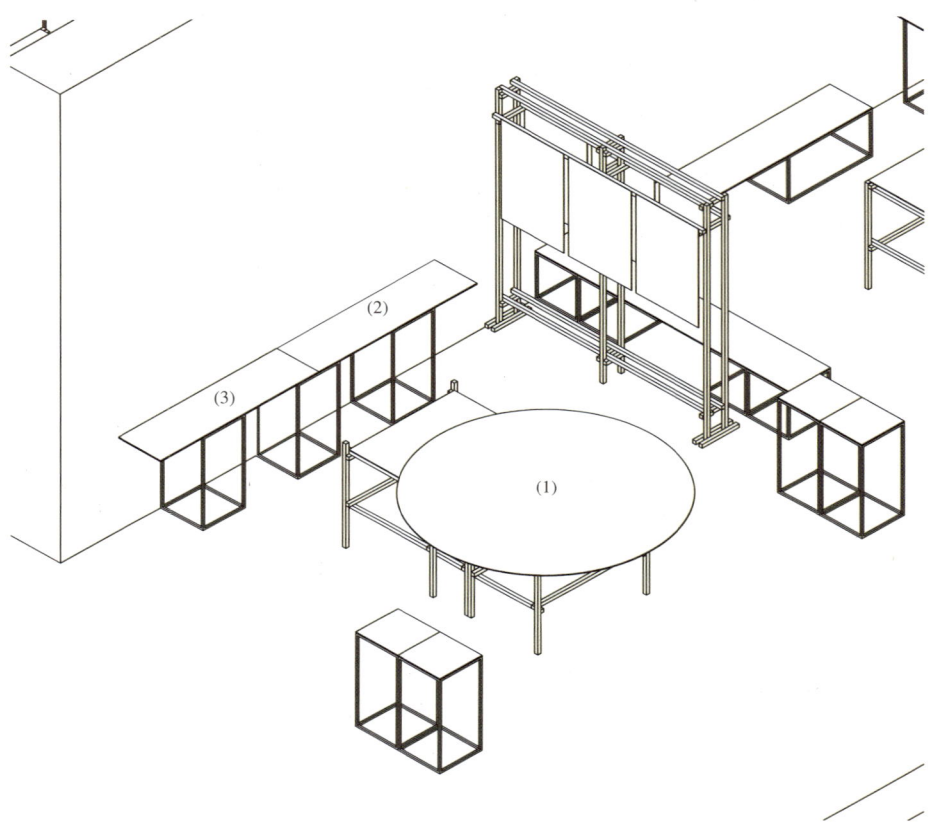

1. Main Model
2. Site Model
3. Structure Model

Exhibition Map — Jiyun Lee

졸업도우미

김진솔
한지원. 조윤서. 정태유. 조남우.
김윤상.

MAIN MODEL

Scale 1:100
Basewood with paper, ⌀ 200 cm

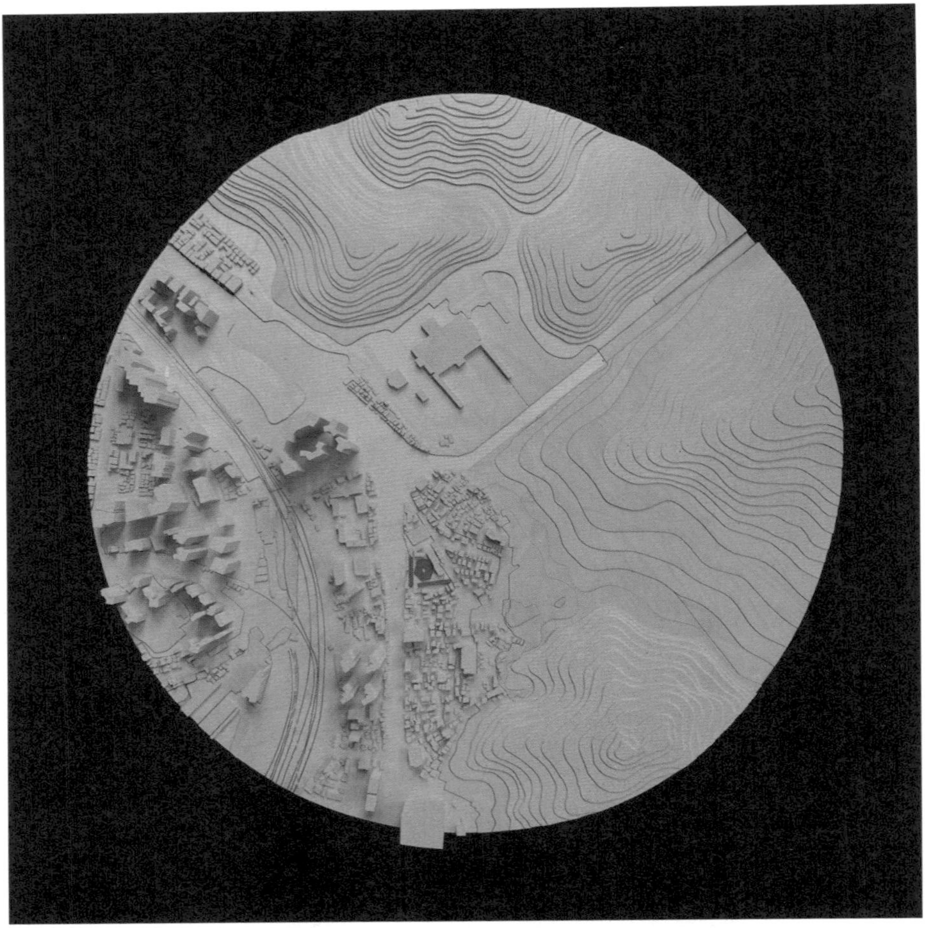

SITE MODEL

Scale 1:1500
Birchwood with 3d print PLA, ⌀ 60 cm

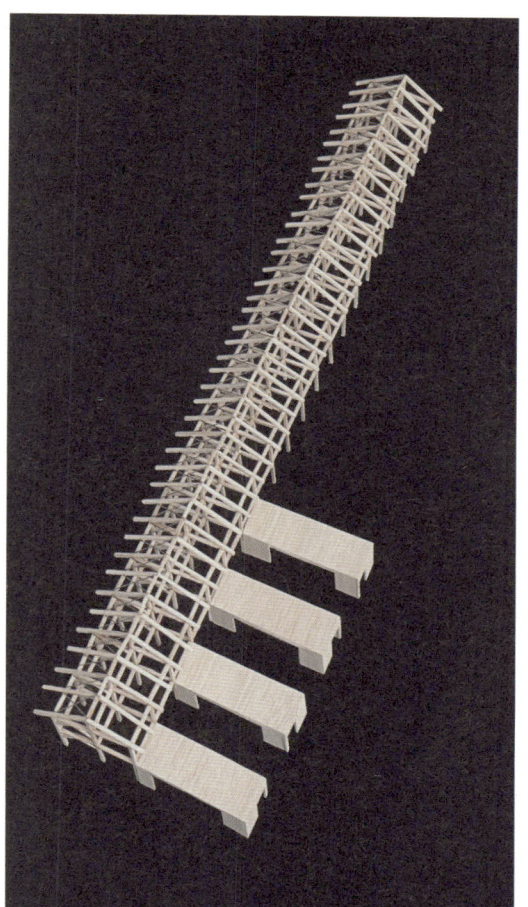

STRUCTURE MODEL

Scale 1:200
Wood and 3d Print PLA, 30 by 30 cm

Tip-Toeing Landscape	Taeyoung Kim Studio
	전희수
	Heesoo Jeon

Preface

Heesoo Jeon

All five of the projects I did in school had Brutalism at the beginning. This was because I felt that it best reflected the time period of labor, war, and class at the beginning of modern architecture. Brutalist architecture, which seems to have crash-landed on another planet and is, in a way, illusory, allowed me to imagine better.

The process of dimensioning and picturing my imagination makes me feel as if 'this nonsense' could be real. And the existence of Brutalist architecture, which would have been someone else's imagination, gives me hope that my imagination will one day exist in a part of the city.

This project imagines a new type of city that reflects the state of the world at this time of transition to a new era. In the unequal land created by capitalism, urbanites begin a small rebellion. Turning on their heels against gravity, they claw their way to a new world.

교내 설계 작품 5개의 프로젝트의 모든 초입엔 브루탈리즘이 있었다. 근대건축의 초입에서 노동, 전쟁, 계급 등의 시대성을 가장 잘 반영했다고 여겼기에 그랬다. 다른 행성에 불시착한 듯한, 어떻게 보면 허상같기도 한 브루탈리즘 건축은 내가 더 좋은 상상을 할 수 있게 하였다.

나의 상상을 치수화하고 이미지화하는 과정은 마치 '이 말도 안되는 상상'이 실재할 수도 있을 것만 같은 생각을 하게 한다. 그리고 누군가의 상상이었을 브루탈리즘 건축이 실재하는 것은, 나의 상상 또한 언젠가 도시 한 부분에 존재할 것이란 희망을 준다.

이 프로젝트는 새로운 시대의 과도기에 선 지금, 이 시대의 모습을 담은 새로운 양식의 도시를 상상으로 그려냈다. 자본주의가 만들어낸 불평등한 땅에서 도시인들은 작은 반항을 시작한다. 중력에 맞서 뒤꿈치를 쳐든 채, 까치발로 새로운 세상을 향해간다.

Main Drawing — Heesoo Jeon

This work seeks to answer the question of polarized cities in a study that begins with megastructures and proposes a new urban paradigm for the future. It may seem a bit bold and sometimes unrealistic, but we hope that this novel attempt, which is new in many ways, will serve as an impetus for human beings to move once again.

이 작품에서는 메가스트럭쳐로부터 시작한 스터디에서 양극화된 도시에 대한 해답을 찾고, 미래에 등장하게 될 새로운 도시 패러다임을 제시한다. 다소 과감하고, 때로는 비현실적으로 비춰질 수 있으나, 여러 방면에서 새로울 이 시도가 정체된 인간을 다시금 움직이게 할 동력이 되길 바란다.

1. Mega Structure
    – Hierarchical urban structure
    – Topographic thermal environment analysis

2. Proposal
    – Development of superblock in Gangnam and heat island
    – A method of circulation; vertical cool island

3. Final Design
    – Mega ventilation structure
    – A curved triangular plane for air circulation
    – Section of the water pump tower
    – Urban transformation plan

The project opens with a study of megastructures. After the industrial revolution, megastructures emerged as saviors, polarizing social classes, and the utopian city of modern architecture was pessimistically viewed. I studied how the paradigm of a new city should be defined, believing that the 'inversion of relationships' would be the answer. I chose Seoul, where rapid urban development has led to polarization of the environment, and Gangnam, where the development of superblocks has left it vulnerable to heat and air, as his site.

The new topography created by the difference in elevation of the buildings turned the blocks into basins, causing hot, stale air to pool in the lower floors, creating an unequal environment. This led to the design of a circulation tower to purify the air in the block and regulate the temperature in the lower floors, forming a new pattern for the city. The tower creates water vapor from the block's groundwater and stored rainwater, which reduces the surrounding temperature. The euphemistically shaped triangular tower allows air to circulate without being trapped.

It also contains temporary housing to accommodate refugees displaced by various disasters. The tower will become a new paradigm of postmodernism and futuristic cities, leading cities to find solutions to environmental inequality.

이 프로젝트는 메가스트럭쳐에 대한 스터디로 막을 연다. 산업혁명 이후, 구세주처럼 등장한 메가스트럭쳐가 사회적 계급을 양극화 하는 현상을 꼬집으며 근대 건축의 유토피아론적 도시를 비관적으로 바라보았다. 새로운 도시의 패러다임은 어떻게 정의되어야 하는가에 대하여 '관계의 도치'가 해답이 될 것으로 보고 스터디를 진행했다. 그에 급격한 도시개발로 환경의 양극화를 불러온 서울의 지역을 꼽았으며, 슈퍼블록 개발로 인해 열과 공기의 취약점에 놓인 강남을 사이트로 설정했다.

건물의 고도차로부터 생겨난 새로운 지형은 블록을 하나의 분지형태로 만들었고, 이는 뜨겁고 탁한 공기가 저층부에 고이게 하여 불평등한 환경이 생겨났다. 이에 도시의 새로운 패턴을 형성하여 블록의 공기를 정화하고 저층부의 온도를 조절하는 순환타워를 디자인하게 되었다. 이 타워는 블록의 지하수와 저장된 빗물로 수증기를 만들어 주변의 온도를 낮춘다.

또한 완곡한 형태의 삼각형 타워를 통해 공기가 갇히지 않고 순환하도록 돕는다. 또한 갖가지 재해로 인해 터전을 잃은 난민을 수용하도록 임시주거지를 담고 있다. 이 타워는 포스트 모더니즘과 미래도시의 새로운 패러다임으로 자리잡아 환경적인 불평등을 개선할 해답으로 도시를 선두하게 될 것이다.

Tip-Toeing Landscape

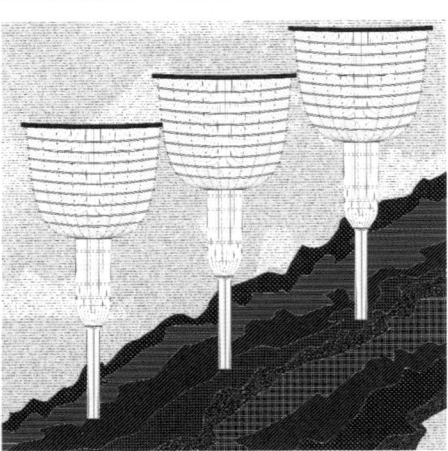

In the transitional period where megastructures and primitive cities coexist, urbanites are becoming disempowered and marginalized. The ideology of capitalism evaluates human leisure activities as meaningless and emphasizes efficiency. With the emergence of skyscrapers, humans look up to an invisible utopia on the top floor, not to the ground on which they stand. Therefore, the urban phenomenon can be divided into three main stages; a spiration, achievement, and the illusion of utopia.

First, the environment in which humans live is the ground, but the rise of skyscrapers has made us lose sight of the ground.

The second stage is the selfishness of urbanites who have achieved their desires to a certain extent. It refers to people who live in well-equipped public environments, such as apartment complexes, who have forgotten their memories of the land and live in fences, throwing down what they don't need.

The third level is the top layer of aspiration. This top layer may not even exist, and when it does, it may not be the utopia we imagine it to be. But the existence of this utopia is what drives us to labor.

These megacities can alienate us, but they can also make us move forward.

메가스트럭쳐와 원시적 도시가 공존하는 과도기에서 도시인이 주체를 상실하고 소외되어가는 상황에 주목하였다. 자본주의라는 이데올로기는 인간의 여가적 행위를 무의미하다고 평가하며 효율만을 중시한다. 마천루의 등장으로 인간은 발디디고 선 땅이 아닌 보이지 않는 꼭대기 층의 유토피아만을 향해 고개를 쳐든다. 이에 도시의 현상을 크게 3단계로 구분하였다. 열망-달성-허구의 유토피아 이렇게 세 단계로 나뉜다.

첫째, 인간이 발 디디고 사는 환경은 땅이지만 마천루의 등장은 땅을 바라보지 못하게 만들었다. 도시인들의 초점은 마천루의 거대한 스케일에 맞춰져 존재의 여부조차 미지한 꼭대기를 바라보게 만든다.

두 번째 단계는 그 욕망을 어느정도 달성한 도시인들의 이기심을 이야기한다. 가령, 아파트 단지 같은 잘 갖춰진 공공환경을 누리고 사는 사람들을 지칭하는데, 이들은 땅에서의 기억을 잊어내고 필요없는 것들을 아래로 내던지며 울타리를 치고 살아간다.

세 번째 단계는 그 열망의 최상층이다. 이 최상층은 존재하는지도 모르며, 실제로 마주했을 때 그 환경이 우리가 상상하던 유토피아가 아닐 수도 있다. 그러나 이 유토피아라는 존재는 인간을 노동하게 하는 요소로 작용한다.

이러한 거대 도시의 모습은 인간을 소외시키기도 하지만 우리를 움직이게 하기도 한다.

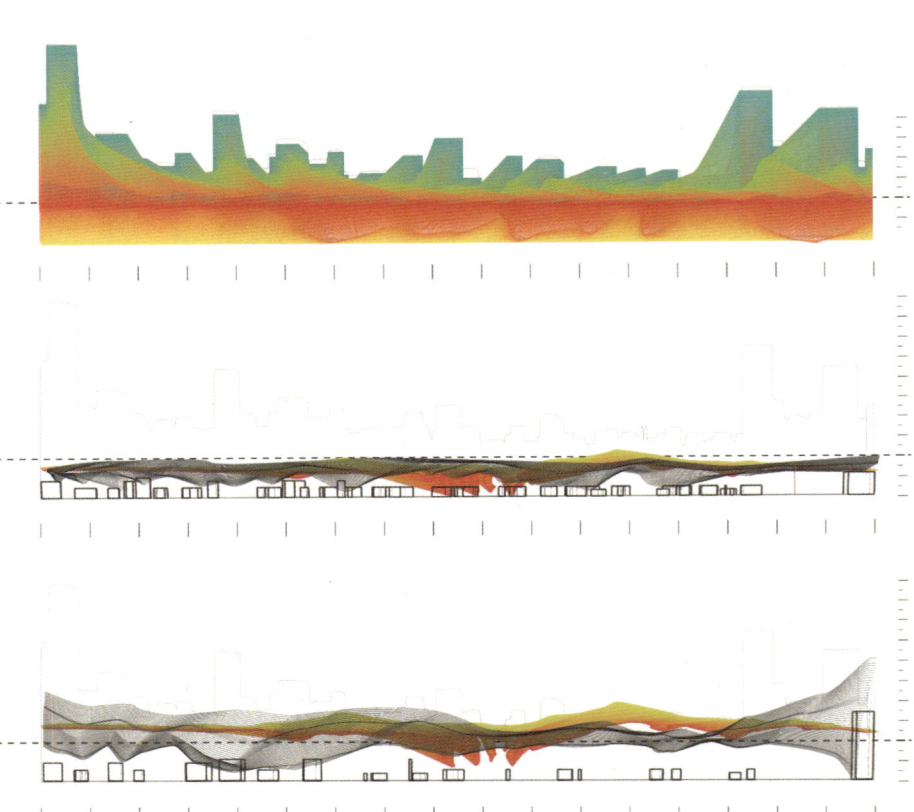

If we look at the background of the emergence of megastructures, we can see that in the 60s, mixed-use commercial buildings appeared, followed by single apartments and then apartment complexes. The type of apartment influenced urban planning, and cities began to be developed in the form of superblocks.

An example is the Yeongdong District Development Plan. As the development progressed, the gap between commercial facilities and multi-family housing widened, creating the appearance that low-rise residences were trapped in a giant wall of buildings. The lower floors below the fifth floor are vulnerable to heat from markets, outdoor units, exhaust heat from automobiles, and radiant heat. The lower floors are also affected by density due to floor-to-ceiling ratio and floor area ratio, making them more vulnerable to air circulation in addition to heat.

The answer to this solid block is a change in microclimate. It is a plan to restore the environment that has been weakened by buildings to buildings again. Using a single block as a reference, I will build a refugee temporary housing tower that will solve environmental problems.

메가스트럭쳐의 등장 배경을 먼저 살펴보면, 60년대에 주상복합형태의 상업시설이 등장했고 그 이후에 단일 아파트, 더 나아가 단지형 아파트가 등장하였다. 아파트의 형태는 도시계획에 영향을 주었고, 이어 도시는 슈퍼블록 형태로 개발이 되기 시작했다.

대표적으로 영동지구 개발계획이 그에 해당된다. 도로에 면한 부분에 상업시설을 놓고 블록 안쪽으로 다세대주택을 배치하였는데, 개발을 거듭할수록 상업시설과 다세대주택의 격차가 벌어지면서 저층주거지가 거대한 건축물 벽에 갇히는듯한 형상을 자아내게 되었다. 5층 이하의 저층부는 시장, 실외기, 자동차에서 나오는 배기열에 의해, 또한 복사열에 의해 열에 취약한 형태에 노출된다. 또한, 건폐율, 용적률에 의해 저층부는 밀도의 영향을 받아 열과 더불어 공기순환에 더욱 취약해진다.

이 단단한 블록을 깨어 줄 틈새는 미기후로부터 시작된다. 개발에 의해 분지형태의 대지가 된 이 부지는 저층고밀주거단지로 배출되지 못한 부산물들이 집적되고, 이들은 도로, 건물틈, 작은 화단 등에서 발생하는 미세한 기후의 전환점에서 배출구를 찾는다. 그 지점에 세워질 기후타워는 꽉막힌 도시에 틈을 만들고 부산물의 출구로 작용한다.

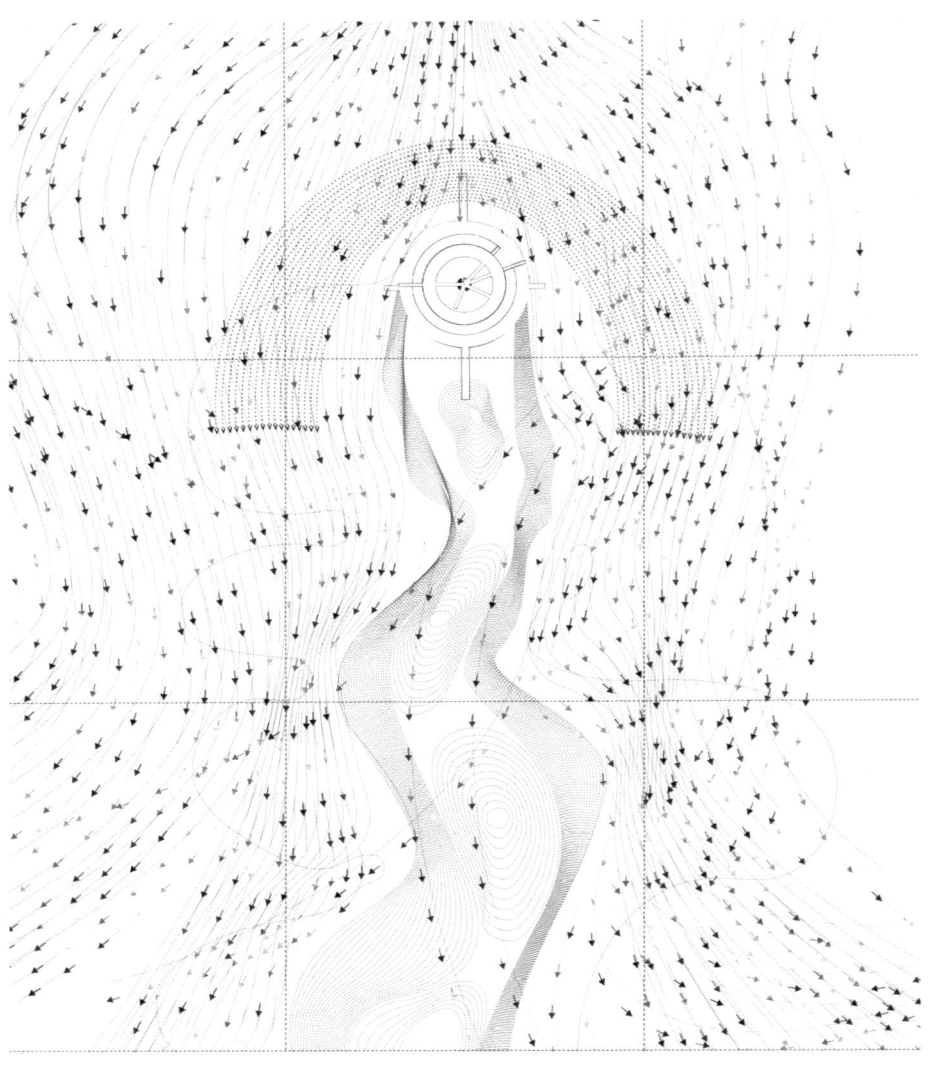

In the transitional period where megastructures and primitive cities coexist, urbanites are becoming disempowered and marginalized. The ideology of capitalism evaluates human leisure activities as meaningless and emphasizes efficiency. With the emergence of skyscrapers, humans look up to an invisible utopia on the top floor, not to the ground on which they stand. Therefore, I categorized the urban phenomenon into three main stages.

First, the environment in which humans live is the ground, but the rise of skyscrapers has made us lose sight of the ground. The second stage is the selfishness of urbanites who have achieved their desires to a certain extent. It refers to people who live in well-equipped public environments, such as apartment complexes, who have forgotten their memories of the land and live in fences, throwing down what they don't need. The third level is the top layer of aspiration. This top layer may not even exist, and when it does, it may not be the utopia we imagine it to be. But the existence of this utopia is what drives us to labor.

These megacities can alienate us, but they can also move us.

메가스트럭쳐와 원시 도시가 공존하는 과도기에서 도시는 무력해지고 소외되고 있다. 자본주의의 이데올로기는 노동을 제외한 인간의 여가활동을 무의미하다고 평가하며, 생산의 효율만을 중시해 왔다. 수직적 구조의 마천루 도시에서 인간은 보이지 않는 꼭대기층의 유토피아를 올려다본다. 인간은 발디디고 서있는 땅보다, 만질 수도, 바라볼 수도 없는 최상층에서 거짓된 소속감을 느낀다. 이에 따라 도시 현상을 크게 세 단계로 분류해본다.

첫번째 단계는 노동자 계급의, 대부분의 인류가 소속하고 있는 땅이다. 두 번째 단계는 어느 정도 욕망을 달성한 중상위 계급의 사회이며, 극한의 집단이기주의를 보여준다. 잘 갖춰진 공공 환경에 살면서 땅에 대한 기억을 잊은 채 울타리 안에서 필요하지 않은 것을 방출하는 사람들을 말한다. 세 번째 단계는 열망의 최상위층 유토피아다. 이 최상층은 어쩌면 존재하지도 않을 수도 있고, 존재하더라도 우리가 상상하는 유토피아가 아닐 수도 있다. 하지만 이 유토피아의 존재는 인간을 노동으로 몰아넣는 원동력이 된다.

이러한 마천루 구조의 도시는 인간을 극한으로 소외시킴과 동시에 인간을 살아가게 한다.

단단한 블록에 공기가 순환할 수 있는 공기쿨링 타워를 두어 일종의 공기 구멍을 만들어 주는 것이 첫 번째 제안이다. 열에 취약한 저층부 블록에 공기 순환 타워를 두고, 이 타워는 도시의 인프라 역할을 도모하며 도시를 순환하게 하는 바퀴의 역할을 한다.

빗물을 받아 안개쿨링을 하며, 주변 환경에 의한 굴뚝 효과로 뜨거운 공기가 거대한 타워를 통해 상층부로 빠져나가게 한다. 이렇게 도시의 모든 수평적 구조를 수직으로 수렴하게 되면서, 이 거대한 타워에는 가로수 경작, 생활수 공급, 전기, 그리고 주거를 위한 공간을 제공한다.

My first suggestion is to place air cooling towers in these solid blocks to allow air to circulate, creating a kind of air pocket. By placing air circulation towers in the low-rise blocks that are vulnerable to heat, these towers serve as the city's infrastructure and act as wheels to circulate the city.

They catch rainwater for fog cooling, and the chimney effect of the surrounding environment allows hot air to escape through the massive towers to the upper floors. This vertical convergence of all the horizontal structures of the city provides space for tree cultivation, water supply, electricity, and housing.

The internal residential blocks, which were regulated to have a minimum area of 50 square meters at the time of the development of the superblock, were set up as a single lot with an average of 36 lots divided into 15 meter by 1.5 meter areas according to the regulations at the time.

Here, the demolished concrete byproducts form a slope and create a new grid on top of it. The center of the slope takes the form of a sunken boonie, on which a cooling tower with a new proposed residential program is built.

The microclimate-altering triangular structure creates a breezeway in the hard city. The interior of the multi-family residential blocks, which are most vulnerable to heat, are hollowed out at equal intervals to allow wind to escape in the direction of the grid.

Material improvements were also needed for the blocks, which lack greenery and public spaces. With the towers in the center, the vacated plots are filled with greenery to block heat reflected from the ground. The tower collects rainwater and sends it downward, where it is stored below, acting as a cooler that emits cool heat and regulates the temperature of the block.

슈퍼블록 개발 당시 최소 면적 50평 안팎의 규제를 받았던 내부 주거지 블록은 당시 규제에 따라 15미터×1.5미터 면적으로 나뉜 필지를 평균적으로 36개 모은 하나의 필지로 설정했다.

여기서 파괴된 콘크리트 부산물들은 경사지를 형성하여 그 위에 새로운 그리드를 만든다. 경사지 중앙은 움푹 파인 부니의 형태를 띠며, 그 위로 새롭게 제안하는 주거프로그램이 담긴 쿨링타워가 세워진다.

미기후를 변화시키는 완곡한 삼각형 형태의 구조물은 단단한 도시에 바람길을 뚫는다. 가장 열에 취약한 다세대 주거블록 내부를 등간격으로 비워 바람이 그리드 방향으로 빠져나갈 수 있도록 비워둔다.

녹지와 공공 환경이 부족한 블록에 재료의 개선도 필요했다. 중앙부에 타워를 두고 비워둔 필지를 녹지로 채워 대지로부터 반사되는 열을 차단한다. 타워에서 빗물을 받아 아래로 내려보내면 아래 저장되어 시원한 열을 뿜는 쿨러의 역할을 하며 블록의 온도를 조절하게 된다.

Tip-Toeing Landscape

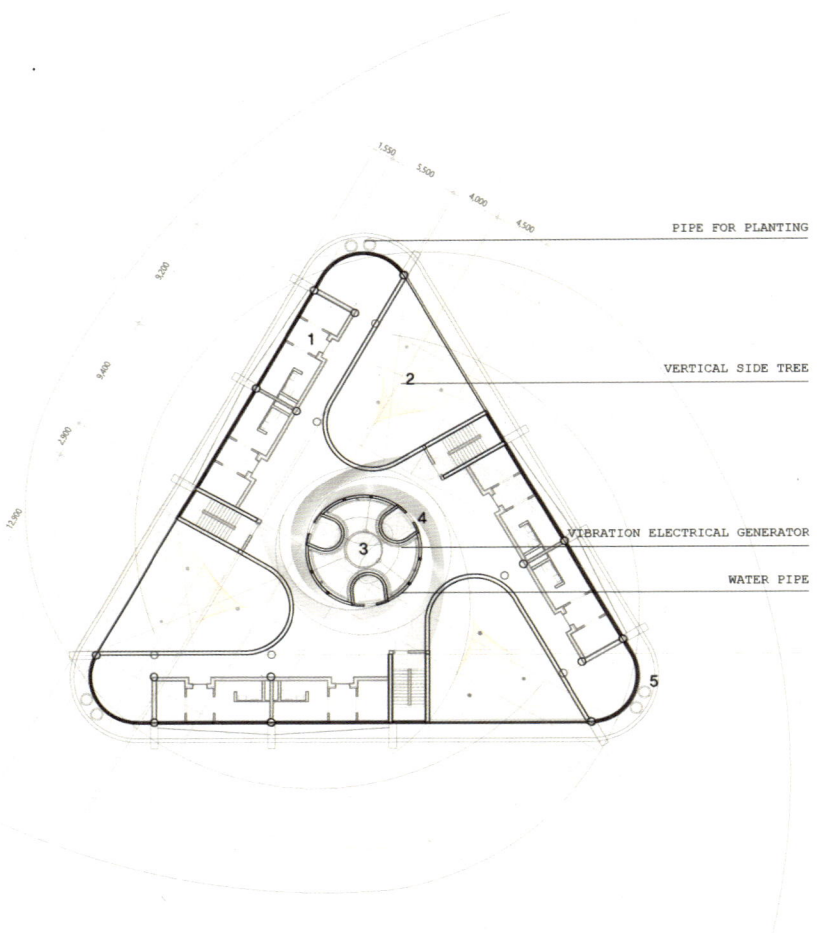

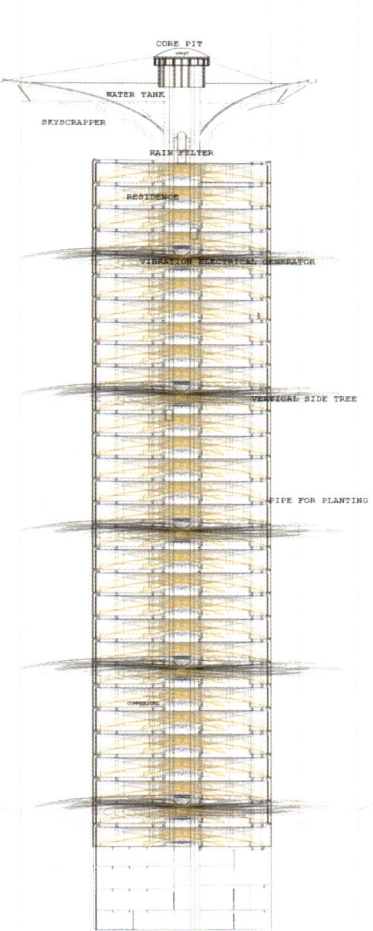

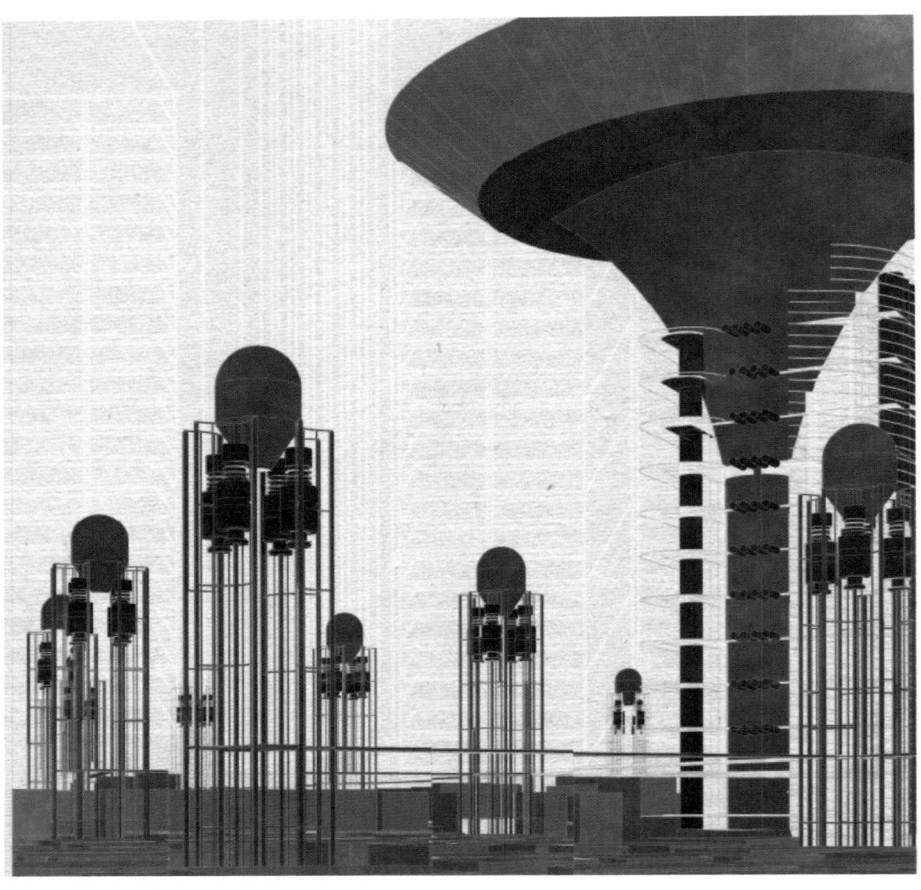

Perspective Drawing

Heesoo Jeon

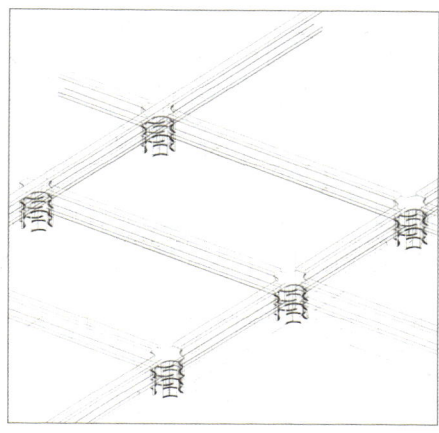
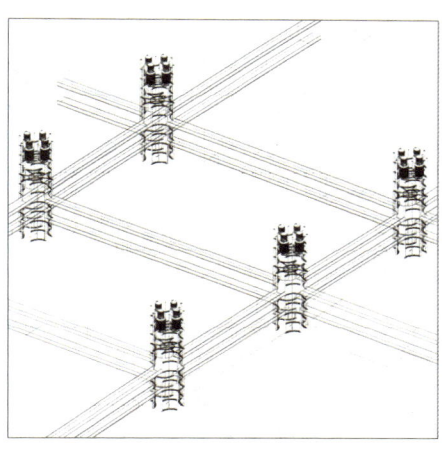
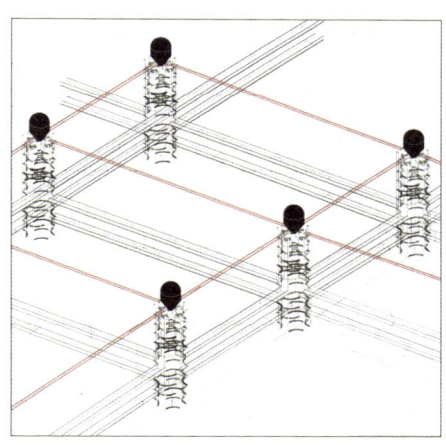

Tip-Toeing Landscape

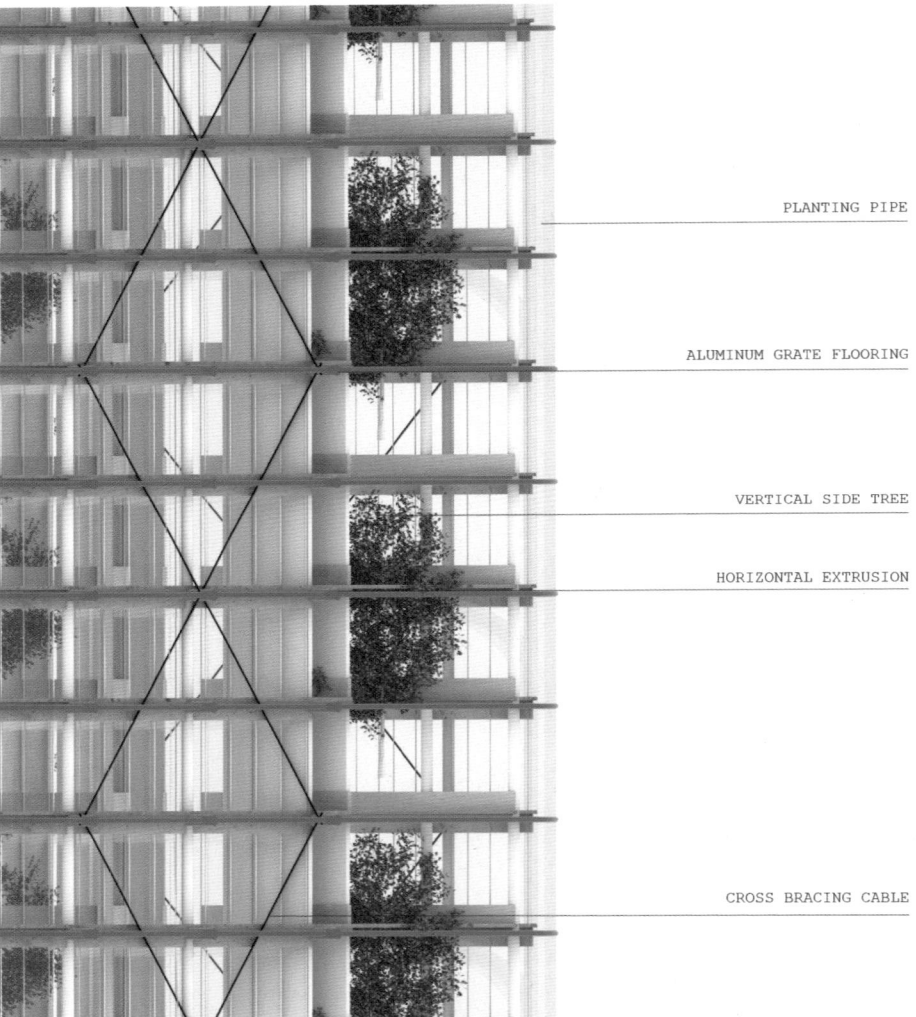

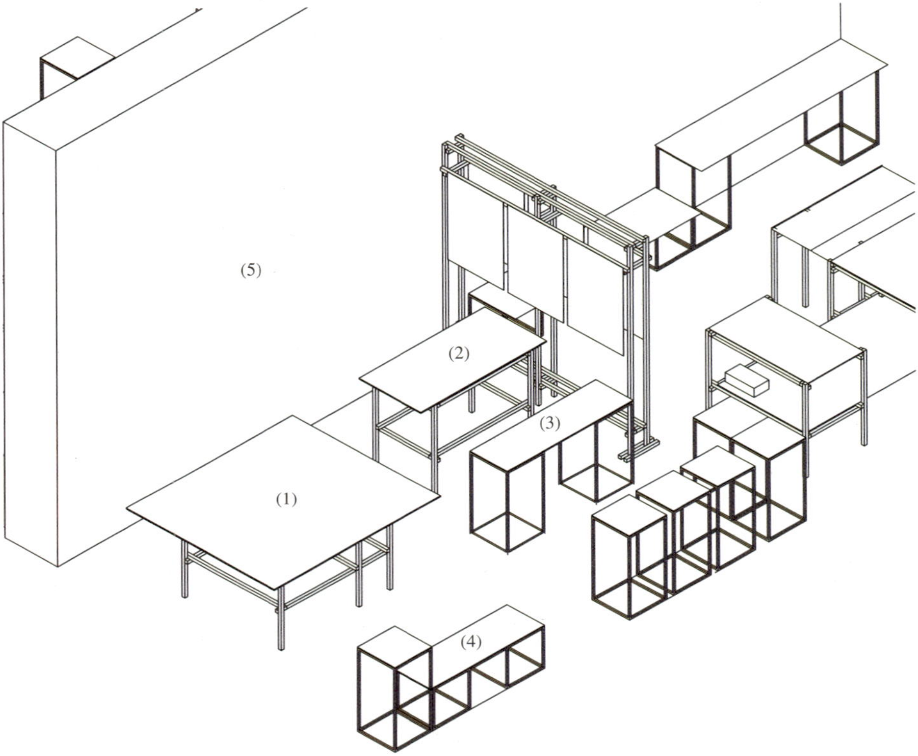

Tip-Toeing Landscape

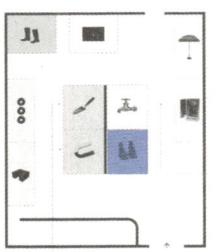

1. Main Model
2. Section Model
3. Structure Model
4. Site Model
5. Video

| 졸업도우미 | 김지우 박송이 |
| | 이선재 이지수 박민서 이영서 |

MAIN MODEL

Scale 1:700
Acrylic, 80 by 80 cm

Tip-Toeing Landscape

SITE MODEL

Scale 1:2000
Paper & 3D print, 60 by 120 cm

STRUCTURE MODEL

Scale 1:100
Paper & Woodrock, 45 by 90 cm

Dance Rock-et	Taeyoung Kim Studio
	조은솔 Eunsol Jo

316   Dance Rock-et

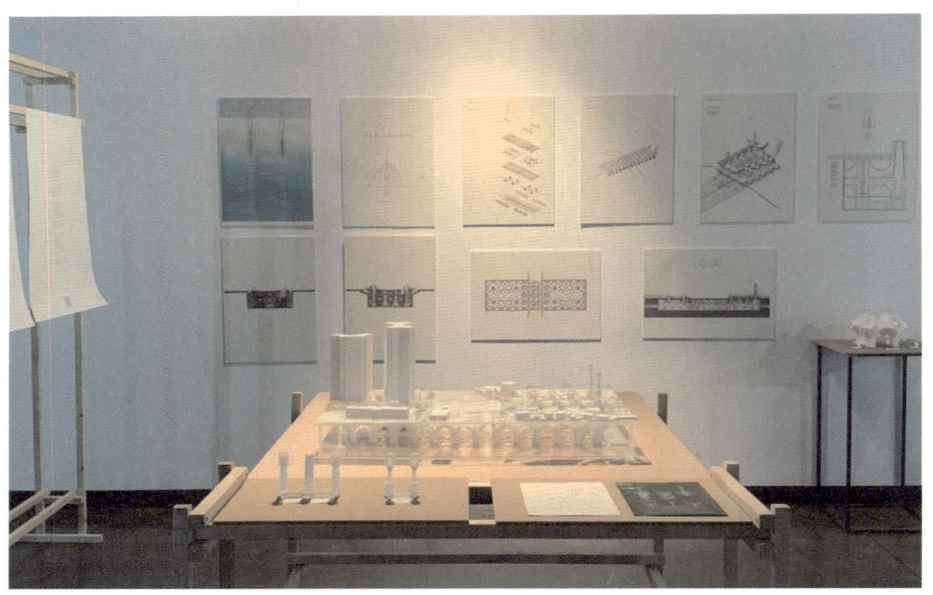

The exhibition is designed with the audience in mind, showcasing the main models and flat works on the left, while on the right, there are videos allowing visitors to experience the sensory aspects of the proposed vibrations and study models. The goal was to present experimental works in a way that makes them easily understandable, illustrating the potential of universal imagination that has been a source of contemplation throughout my five-year journey.

I am grateful to my friends, teachers, and family who have been by my side from the beginning to the end of these five years. I remember the people who filled in the fragments of every stage in the process over the course of five years.

전시는 관람객을 기준으로 왼쪽에 메인 모델과 평면 작업을 전시하고 오른쪽은 작업에서 제안하는 진동의 감각을 체험하는 영상과 스터디 모델을 전시한다.

5년의 시작부터 마무리까지 옆에 있어준 친구들 선생님 그리고 가족에게 감사하다. 5년의 모든 과정의 단편을 채워준 사람들을 기억한다.

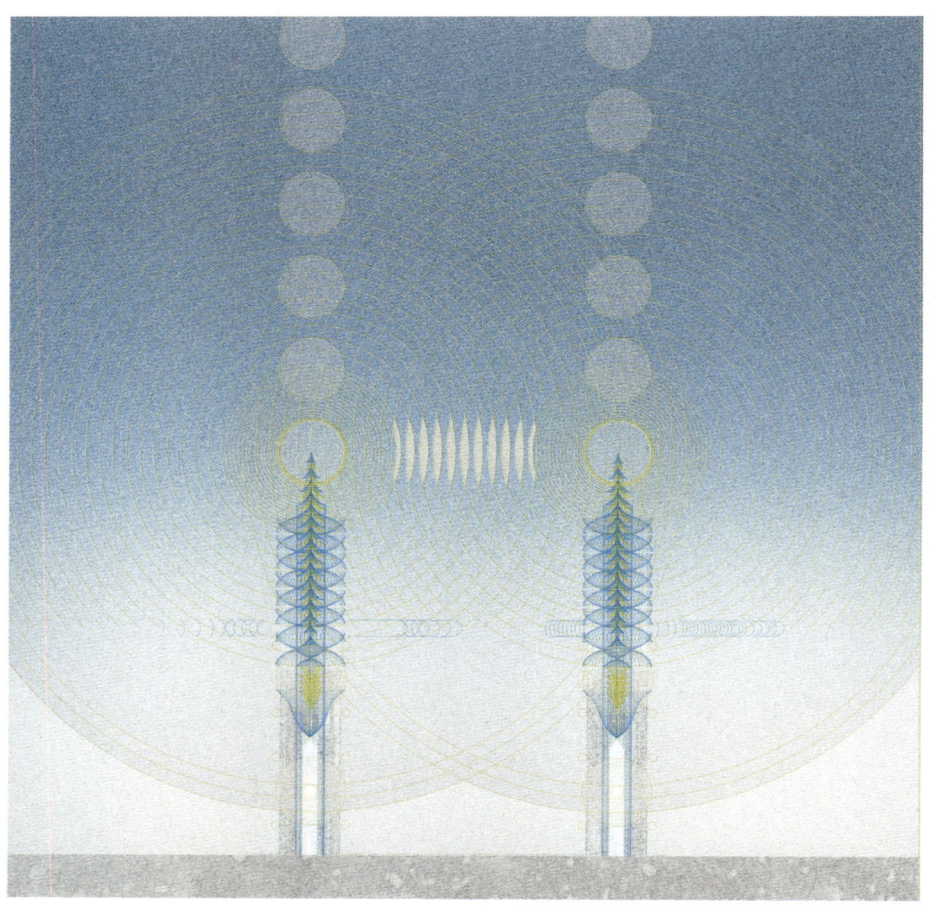

Dance Rock-et

Main Drawing Eunsol Jo

This project introduces an innovative approach to history, generating and harnessing energy from subway vibrations and pedestrians' footsteps within a triangular area comprising long-unused transit corridors. Transforming the sensation of travel through vibrations and wind into a new form of production and labor, this initiative envisions an active public station. It involves the analysis of air flow, vibration, and sound, converting these elements into a spatial experience crafted by individuals. This endeavor not only provides an opportunity for inclusive and universal imagination but also assigns a fresh role to Seoul's subway stations, redefining them beyond their conventional functional purposes. By imbuing these spaces with environmental and spatial experiential significance, we propose a novel perspective on public space in the city.

본 프로젝트는 환승의 용도로만 이루어져 긴 유휴 환승통로를 가진 삼각지역에 지하철의 진동과 사람들의 발걸음으로 에너지를 만들고 이를 경험하는 새로운 형식의 역사를 제안한다. 진동과 바람이 주는 여행의 감각은 에너지를 만들며 새로운 생산, 노동이 된다. 공기, 진동, 소리의 흐름을 분석하여 이를 공간의 경험으로 전환시키고 개개인이 모여 만들어내는 능동적인 공공역사는 공평하고 보편적인 상상의 기회를 제공한다. 기능적인 목적으로만 설계되어 온 지하철 역을 대상으로, 기존에 고려되지 않았던 환경적 공간 경험적 역할을 부여하여 서울에서 제시해야 하는 새로운 공공공간의 역할을 제안한다.

1. 움직이는 인간의 이야기

2. 움직임, 상상, 생산환동

3. 보편적인 생산활동으로서의 상상

4. 환승역 삼각지

5. 지하철을 통과하는 진동 바람의 분석

6. 공간의 모듈화

7. 평면도

8. 단면도

9. 사람과 지하철이 생산하는 에너지 플랫폼

10. 단면 투시도

11. 인프라와 경험공간의 일체화

12. 모델

Our inquiry commenced with a contemplation of how individuals in today's society fulfill their innate need for imagination. In the course of our investigation, we noted a contemporary trend where traditional imaginative pursuits are supplanted by activities like travel or clubbing, catering to the modern appetite for sensory experiences. In our quest for a universally imagined space, we present an innovative public area that blends the vibrational essence of a subway station with insights drawn from the sensations associated with travel and clubbing. This envisioned space is intricately crafted, featuring modules that revolve around the dynamics of wind and vibration. Each module is purposefully designed to serve a specific function, encompassing elements such as movement, ventilation, and energy production. This meticulous design ensures a harmonious integration of form and function within the space.

인간의 상상에 대한 기본적인 욕구를 현대사회에서 사람들은 어떻게 해소하고 있는지에 대한 물음을 시작으로 상상의 감각을 찾아나서는 현대의 우리를 관찰했다. 움직임의 감각을 대체하는 여행 또는 클럽 등이 현대인의 갈증을 해소한다. 보편적인 상상이 가능한 공간을 찾기 위해 여행의 감각, 클럽이라는 공간이 주는 감각에 대한 리서치를 기반으로 진동을 만드는 지하철 역사와 결합한 새로운 공공공간을 제안한다. 공간은 바람과 진동의 움직임을 기반으로 모듈을 디자인 하고 각각의 모듈이 이동, 환기, 에너지 생산의 기능을 담당한다.

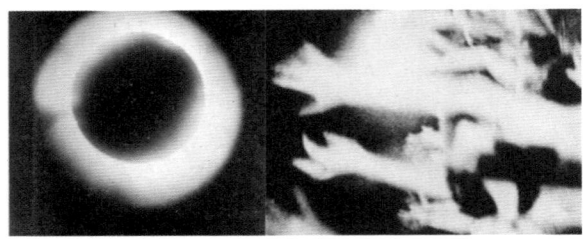

Travel
Station
Dreaming

'Imagination' should be equal to everyone as a universal 'productive activity' rather than a 'surplus' of privileged people.

하나의 공간에 존재하는 나는, 끊임없이 꿈꾸는 인간이어야 하며, '상상'은 특권을 가진 사람들의 '잉여'가 아닌 보편적인 '생산적인 활동'으로서 모두에게 평등해야한다.

움직이는 인간의 이야기

인간은 원래 움직이는 동물이었다. 호기심을
가진 인간은 상상하며 계속 걸었다.
움직임 자체가 생산이었다. 꿈을 꾸고 다음을
기대하면서. 호기심을 잃은 몇몇은
정착했고 그것을 책임이라고 이름지었으며
서로를 나누는 차별로 이어졌다. 상상의
감각을 그리워하는 우리 현대인은 그 감각을
대체할 '여행'을 가거나, 파티에 간다.
움직임을 멈춘 인간은 가보지도 않은 장소를
그리워한다.

현대사회에서 모두가 평등하게 상상의
기회를 가질 수 있을까.

존재하는 나는, 끊임없이 꿈꾸는 인간이어야
하며, '상상'은 특권을 가진 사람들의
'잉여'가 아닌 보편적인 '생산적인
활동'으로서 모두에게 평등해야한다.

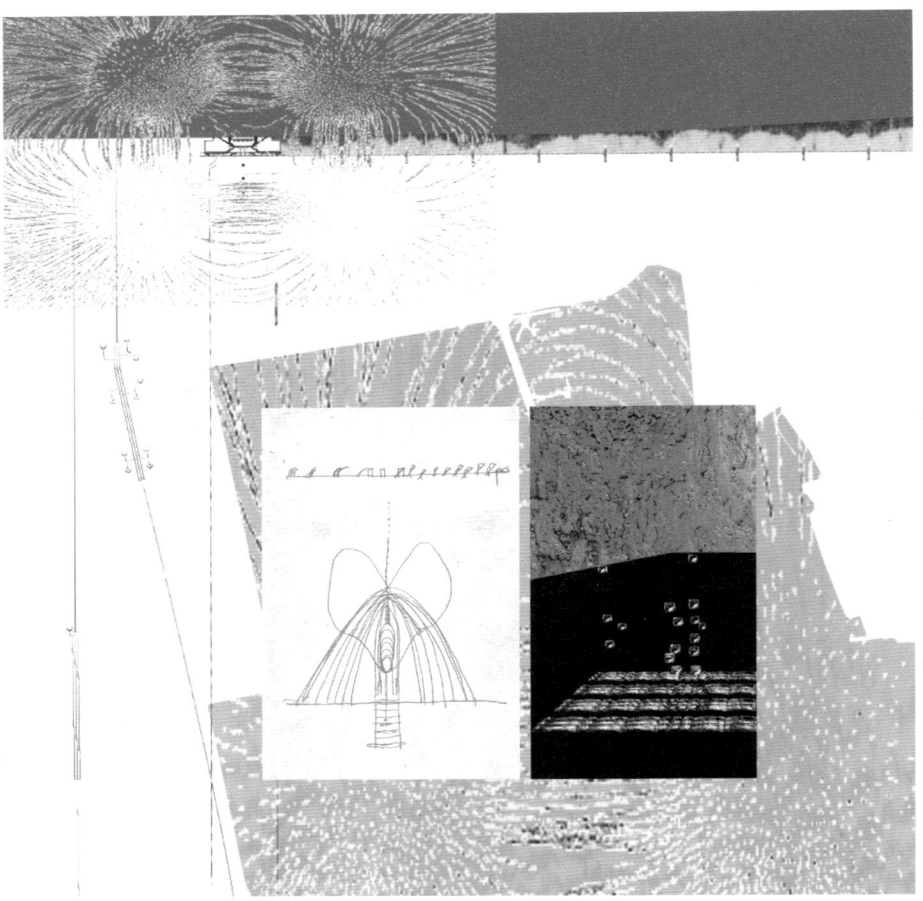

The story of humans in motion

Humans were originally animals in motion. Curious, we walked and walked and walked. Movement itself was production. Dreaming and looking forward to the next. Some who lost their curiosity settled down and named it responsibility, which led to discrimination that divided us from one another. We moderns, who miss the sensation of imagination, go on "trips" to replace it, or go to parties. When we stop moving, we long for places we've never been.

Can everyone have an equal opportunity to imagine in modern society?

The existence of me must be that of a human being who is constantly dreaming, and 'imagination' must be equal to all as a universal 'productive activity', not a 'surplus' of the privileged.

Seoul Geological Metro Map

Vibration Scape of Samgakji

Pedestrian Steps

Overground transportation

Underground transportation

Samgakji Station Section Plan

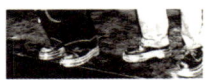

Liminal Space and Clubbing District

Site : This area became louder during the night. There are Clubbing district near Samgakji Station. On the otherhand station itself goes to sleep since the metro stops till the morning 5:40.
It has long liminal spaces caused by transfering between two lines. What if this station never sleeps.

Can we collect the vibration and make it into energy?

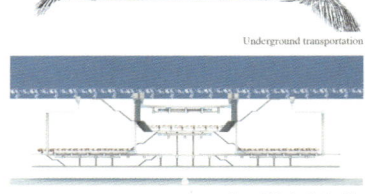

삼각지역은 환승 목적에 따라 비어있는 긴 통로를 가지고 있다. 진동의 형태로 그려진 지하의 공간은 새로운 공공 역사의 가능성을 묘사한다. 환승의 목적으로 만 사용되고 있는 긴 유휴공간을 사람들이 찾아와 생성하는 발걸음으로, 지하철이 지나며 만드는 진동으로 에너지를 생산하는 역사를 상상한다. 진동이 그려내는 지하의 스케이프는 아름답다. 전쟁기념관의 공원의 연장선에서 지하의 공간은 스케이트 공원, 콘서트홀 등의 프로그램을 가진다.

Liminal Space, Samgakji Station

Seoul Vibration Level

Losts of steps and rail line makes different level of vibrations. Liminal spaces underground could be a potential space of collecting all vibration we make in daily life.

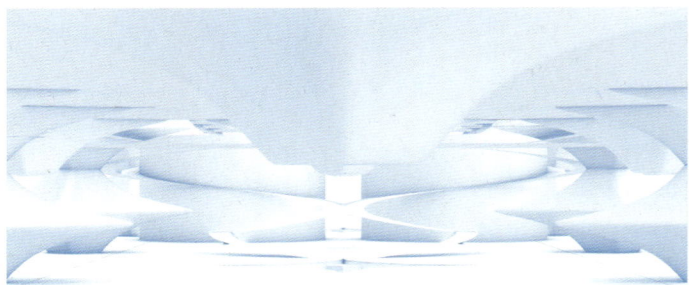

Visualization, Vibration Collecting Metro Station

Samgak-ji Station boasts an extensive passageway, traditionally left vacant for transit purposes. Envisioning a new public history, this underground space, shaped by vibrations, explores the potential for generating energy from the subway's passing vibrations. People traverse the area, transforming long idle spaces intended solely for transit into active contributors to this unique historical narrative. The subterranean landscape, animated by the rhythmic vibrations, exudes a captivating beauty. Functionally extending from the War Memorial's park, this underground expanse accommodates diverse programs, including skate parks, concert halls, and more. It represents a departure from the conventional use of underground spaces, offering a dynamic and multifaceted environment for cultural and recreational activities.

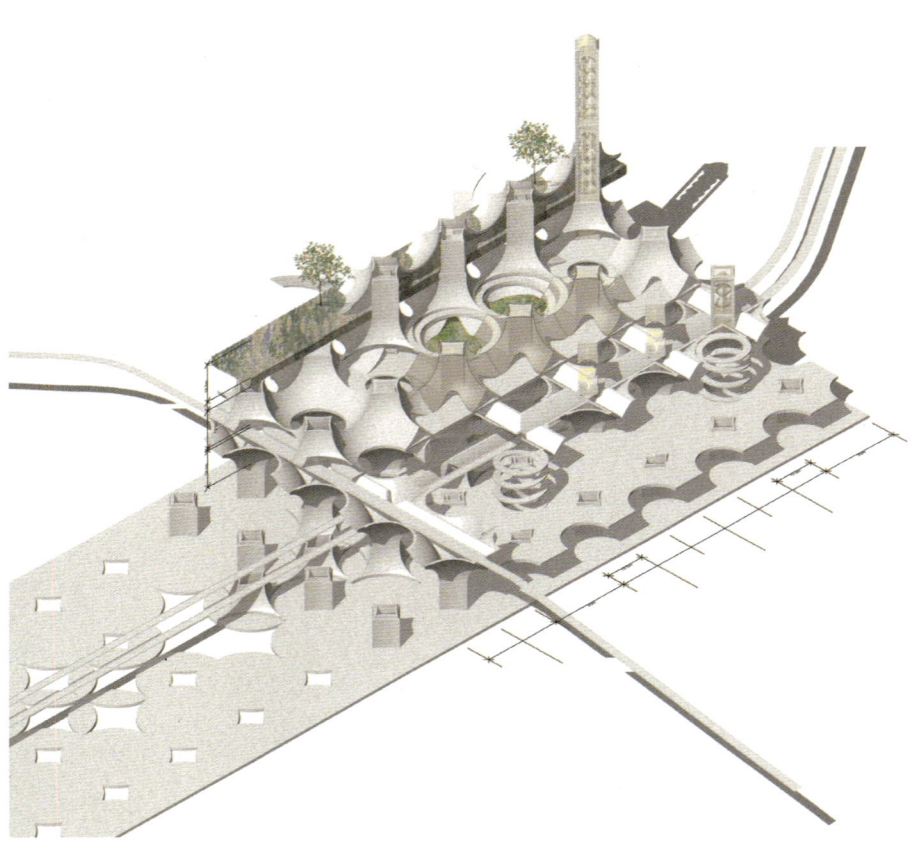

Mass Study — Eunsol Jo

The space is meticulously designed to convey wind and vibration movements, with each module assigned specific functions related to movement, ventilation, and energy production. In conventional subway designs, there is a clear separation between the areas designated for ventilation machinery and those meant for pedestrian traffic. The diamond-shaped module introduced here not only incorporates the vibrations from the subway but also integrates the rhythmic footsteps of people, transforming the entire system into a unified and experiential facility.

공간은 바람과 진동의 움직임을 잘 전달할 수 있는 모듈을 디자인 하고 각각의 모듈이 이동, 환기, 에너지 생산의 기능을 담당한다. 보통 지하철은 환기 기계설비를 위한 공간과 사람이 지나는 공간을 분리한다. 지하철이 그리는 진동에 사람들의 발걸음을 더하는 다이아몬드 형태의 모듈은 경험과 더불어 설비 시스템을 하나로 통합시킨다.

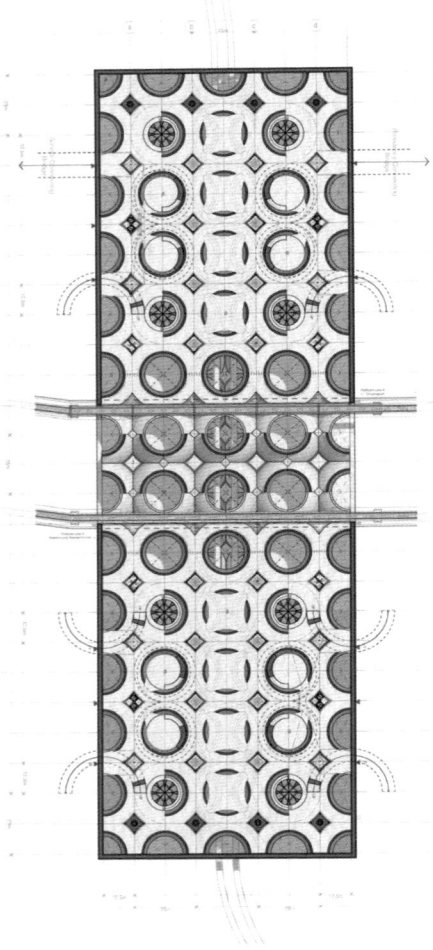

-1 Floor Plan

The layout, which is drawn as a list of modules, has a system that can be applied to other Seoul stations depending on the size of the subway.

모듈의 나열로 그려지는 배치는 지하철의 크기에 따라 다른 서울의 역사에도 응용될수 있는 시스템을 가지고 있다

-2 Floor Plan

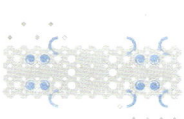

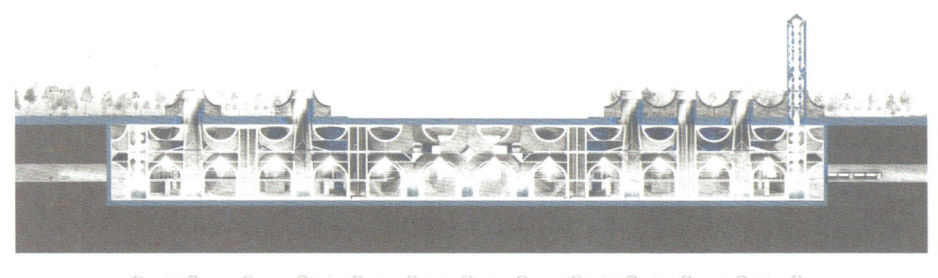

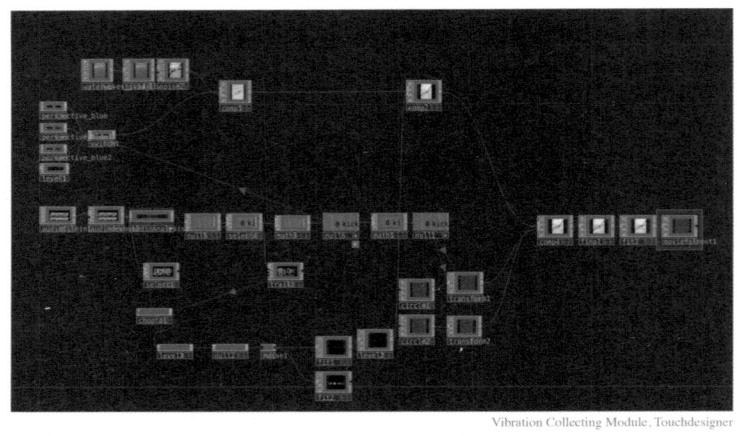

Vibration Collecting Module, Touchdesigner

| Sound of Metro | Divide sound by the vibration | Different kicks created | Transforming kicks by the Vibration | Makes Visual effect in every Kicks accoring to different vibration level | Vibration becomes Energy |

Dancing Metro, Visualization

Vibration Collecting Module Video

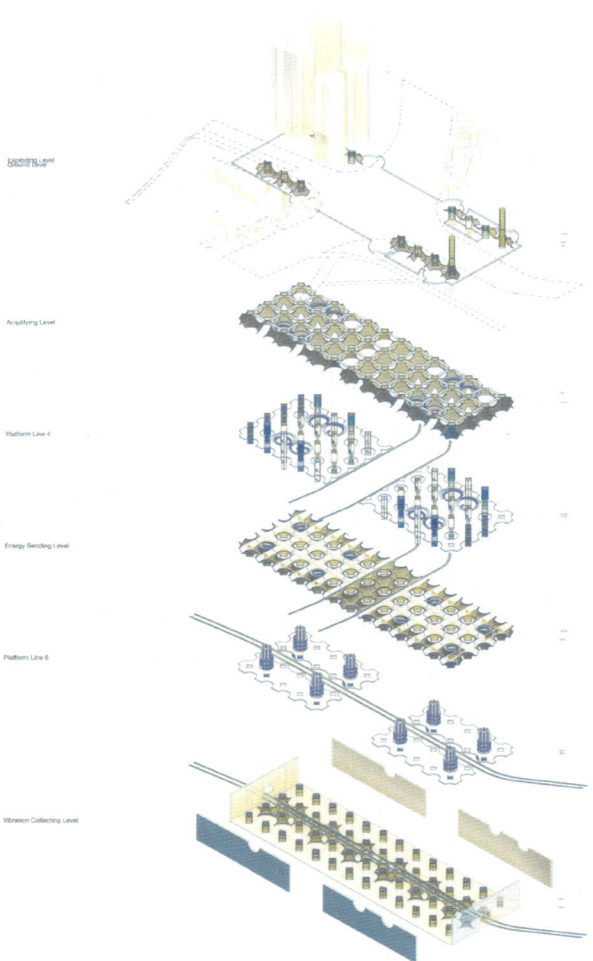

Each module performs functions such as ventilation, energy production, and transportation, and people move through the space between them. All movements are transmitted and stored through the modules, forming an integrated system.

각각의 모듈은 환기, 에너지생산, 이동 등을 담당하며 그 사이공간을 사람들이 오간다. 모든 움직임들은 모듈을 통해 전달되고 저장되며 하나의 통합시스템을 형성한다.

Dance Rock-et

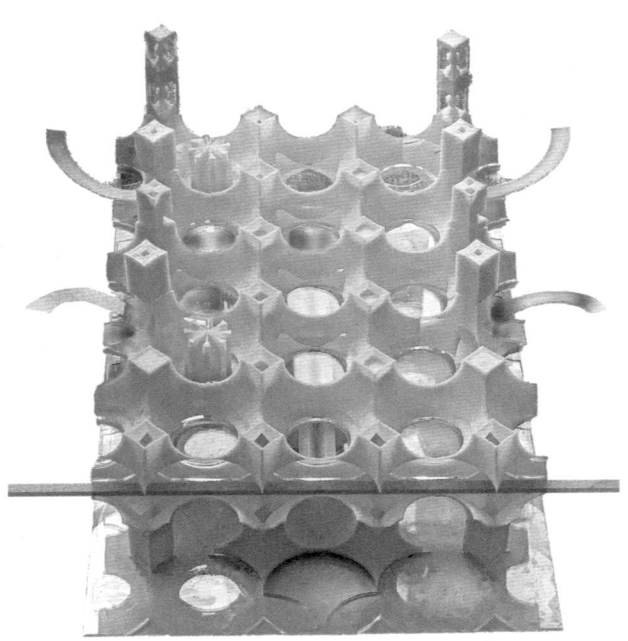

STUDY MODEL

Scale 1:500
3D print PLA, 60 by 30 by 15 cm

STRUCTURE MODEL

Scale 1:100
3D print PLA, 40 by 40 by 40 cm

부록
Appendix

Graduation Exhibition

2022. 12. 22 – 12. 27

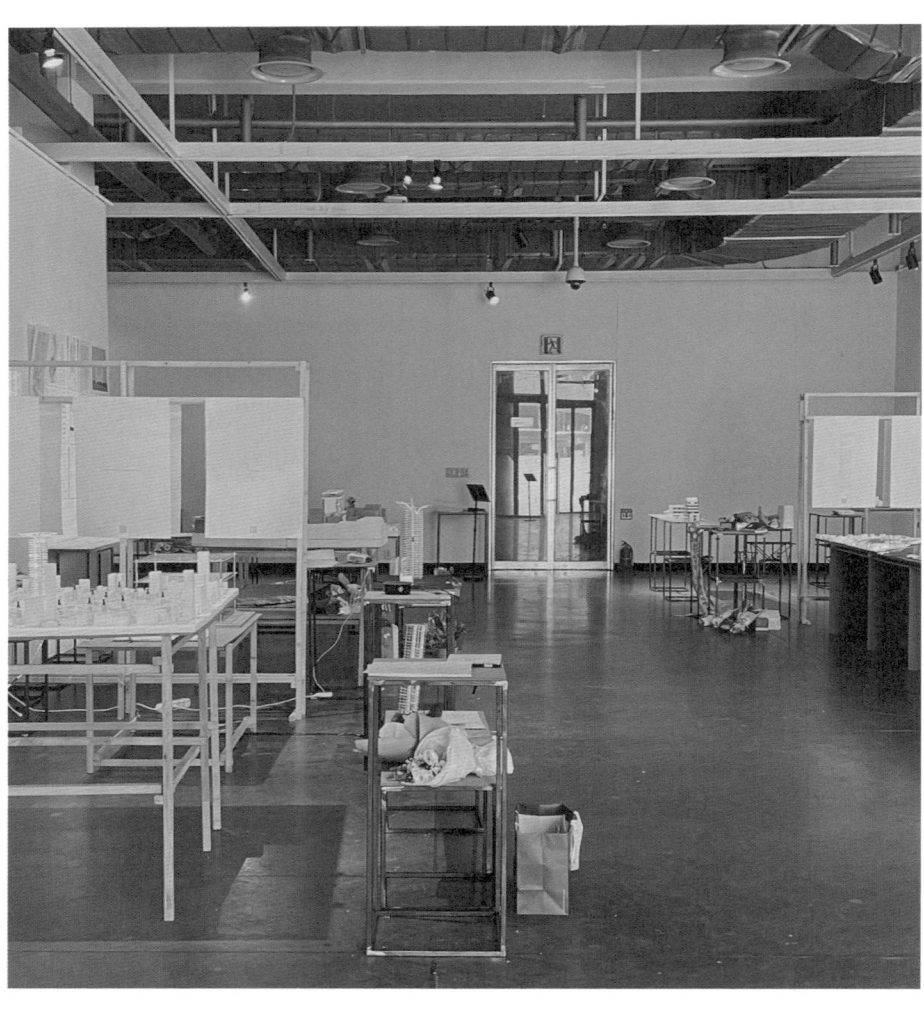

Exhibition Archives

Directors' note

This journal chapter will be the final one from orbit. For five years, we have been off of the planet, free from the tugging bond of gravity. We have flown like Superman, flipped and twisted like a gymnast, and even whacked our head on a handrail once or twice. None other planet has blue water and white clouds, covering colorful landmasses filled with thriving, beautiful, living things. We obsesrve our planet day by day. Big forest, big ocean, and creature in between. They build, eat, talk, and walk. Every phenomenon has its own connection. We really can't take that too lightly.

The universe refers to infinite time and space, and to everything contained. Under this context, we focused on ways to ensure continuity of the Earth from a universal perspective, and of the daily lives for human beings who live on it. It is an exercise of discovering possibilities in a periodic condition of increasing physical, social and environmental limitation.

We are heading back home.
When the Shuttle re-enters our beautiful blue blanket of atmosphere, the reality of gravity may jump on our body. We may stagger, we may stumble, we may even fall down. No matter what happens, we will have the biggest smile that we can muster. Now we face our eath, sway in the breeze, listen to the echo back form the valley.

First, last, and many more to come.
This is a five-year journal, also our coordinate of the cornerstone, implying each new chapter of us 9.

전시 기획 의도

5년간 중력의 구속에서 벗어나
우주를 유영했습니다. 슈퍼맨처럼 날고,
운동선수처럼 몸을 비틀고 뒤집으며
그러다 가끔은 머리를 부딪히기도 했습니다.
먼발치 떨어져 바라본 지구는 충분히
아름답고 고귀합니다. 아주 많은 숲과
큰 바다가 있습니다. 생명체도 삽니다.
집을 짓고, 대화를 나누고, 걷습니다.

모든 현상에는 저만의 관계가 흐릅니다.
우리는 천체 속의 지구, 지구를 사는
사람들의 관계와 일상이 지속성을 갖도록
하는 방법에 천착했습니다. 증가하는
물리적, 사회적, 환경적 한계 앞에서도
가능성을 발견하는 연습입니다.

셔틀이 지구로 돌아갑니다.
대기권으로 진입하는 순간부터, 중력의
현실이 몸 위로 뛰어올라 한동안은 비틀거릴
수도, 넘어질 수도 있습니다. 여전히
기쁜 이유는, 오래도록 지켜만 봤던 별에
다시 속하기 때문입니다. 땅을 직접 만지고,
바람을 맞고, 소리도 들을 수 있습니다.

무사착륙의 첫발을 내딛습니다.
9인의 5년 기록이자 새로운 시작을 정초하는
자리가 될 것입니다.

Graduation Exhibition

2022. 12. 22 – 12. 27

Back to the Earth

한국예술종합학교
건축과 2022 졸업전시

2022.12.22 - 12.27

한국예술종합학교 석관캠퍼스 본관갤러리
월-토 09:00-18:00 / 일 10:00-16:00
전시 오프닝 12.22, 18:00

강하림 김보경 김서진 문희윤 신지숭
이연우 이은우 이지률 진희수 조은슬

1. Exhibition Poster, A2

349 전시 기록 Exhibition Archives

2. Postcard, 3 types

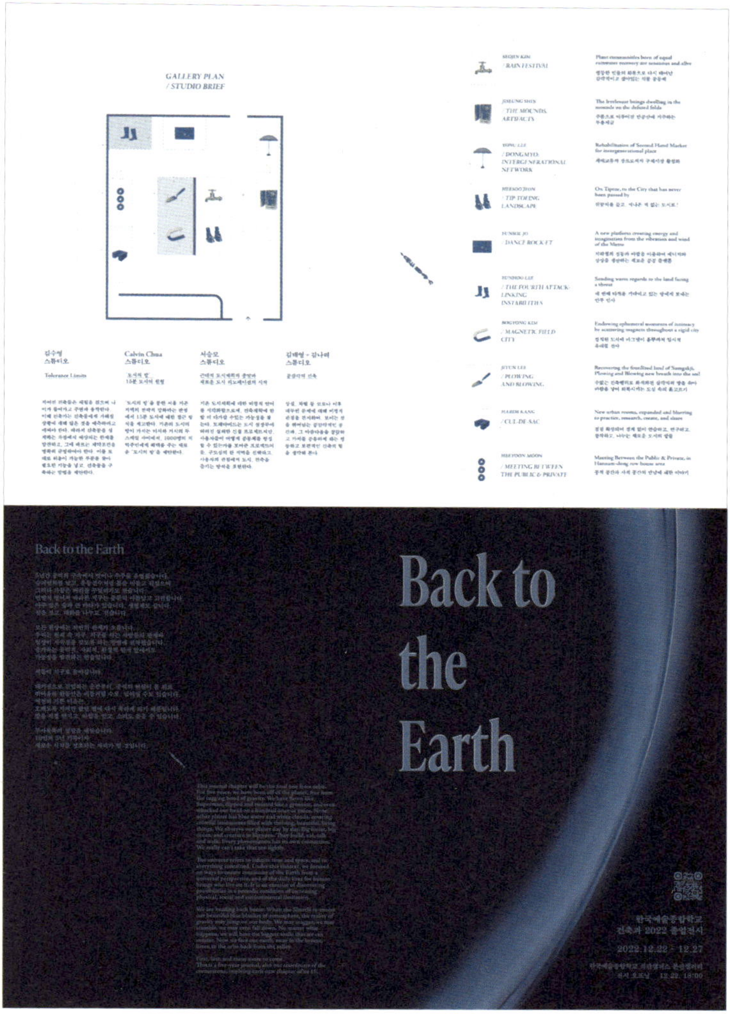

3. Leaflet, A4

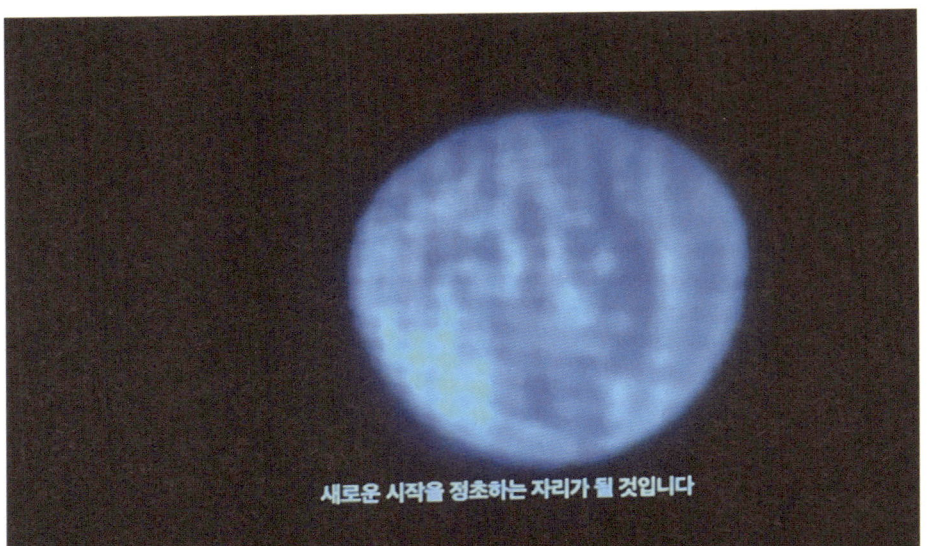

4. Main Video

5. Interview Video

Video Making

We produced two videos for the exhibition.

We created a 10-minute video to accompany the exhibition's theme, Back to the Earth, and to showcase the exhibition's curatorial intentions of the students who participated in the graduation exhibition, which was shown at the exhibition opening and during the exhibition.

A separate video was created featuring interviews with the graduating studio teachers. The interviews included words of wisdom for the graduating students and reflections on the past year. The interviews were shown at the opening of the exhibition.

영상 제작

전시를 위해 두 개의 영상을 제작했다.

전시의 주제였던 Back to the Earth 와 함께 졸업 전시에 참여한 학생들의 전시 기획 의도를 보여줄 수 있는 10분 가량의 영상을 제작했고, 전시 오프닝과 전시기간 중에 상영되었다.

졸업 스튜디오 선생님들의 인터뷰가 담긴 영상을 따로 제작했다. 인터뷰 영상에는 졸업을 앞둔 학생들에게 해주고 싶은 말과, 지난 1년을 돌아보는 내용이 담겨있다. 인터뷰 영상은 전시 오프닝 때에 상영되었다.

Graduation Exhibition

2022. 12. 22 – 12. 27

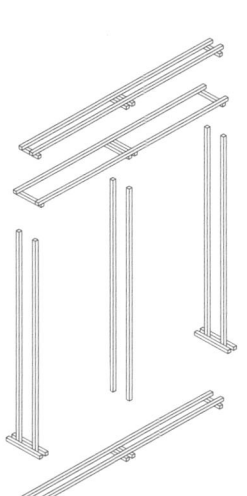
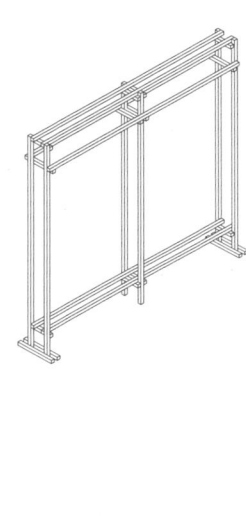
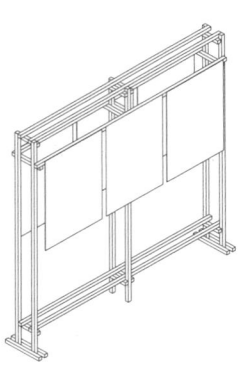

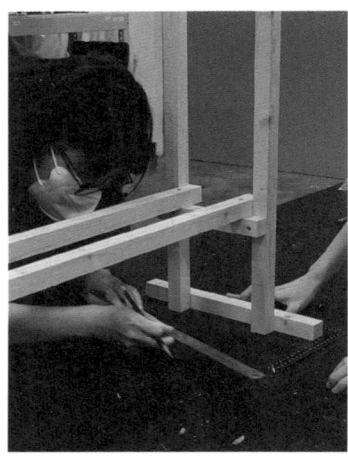

6. Partition

Exhibition Archives

Partition Making

We maked the customized partitions for the exhibition by weaving a 30×30 square of wood plank. The width was based on the standard of placing an A1 portrait-sized paper horizontally, and the height was based on the standard of eye level that an A1 portrait-sized paper can be read well while standing in the exhibition hall.

The partitions were designed to be reversible so that the works in the exhibition, which were divided by partitions, could be hung on both sides.

파티션 제작

30×30 각재를 엮어 전시에 쓰일 파티션을 직접 모델링하고 제작했다. 가로 규격은 A1 세로 사이즈의 종이를 가로로 배치할 수 있게 기준을 잡았고, 높이 또한 A1 세로 사이즈의 종이를 전시장에서 서서 잘 읽힐 수 있는 눈높이의 기준에 맞춰 제작했다.

파티션을 기준으로 나뉘어진 전시장의 작품을 각각 양쪽 면에 걸 수 있게 양면 모두 사용가능한 파티션으로 계획하였다.

We started the archive project in March and didn't finish it until the cold winter months. I had an online meeting in Cairns where I agreed to take a break after graduation to work on the book, but then I spent the next few months traveling and doing bits and pieces on my laptop, which pushed the original summer publication date back another half year.

For a long time, I thought a lot about how to organize the portfolios of the nine graduates. Each of them has their own narrative and different colors, so I tried to show them in a few stages, like I did for the graduation exhibition with a loose theme. The most important thing in archiving creative works is to capture the artist's intention, so I thought a lot about the format of the publication. I wanted to showcase everyone's experimental but solid work through layout, Korean fonts, English fonts, and materials of the paper.

I would like to thank professor Don-Son Woo and Taeyoung Kim for lovingly writing about each student's work, tutor Soo Young Kim, Naree Kim, Seung Mo Seo, and Calvin Chua for their design guidance throughout 2022, and professor Taeyoung Kim & RIBA assistant Eunji Kim for their patience and help in putting the book together despite their busy schedules, Designer Jung-ah Lee for designing the cover even while she was in the middle of her graduation examination, Intime for printing and binding, and finally, graduates Harim Kang, Bogyong Kim, Seojin Kim, Jiseung Shin, Yonu Lee, Eunhoo Lee, Heesoo Jeon, and Eunsol Jo for capturing their wonderful work.

Jiyun Lee
Editor, 2023 Graduate of Department of Architecture

맺으며 이지윤

3월에 시작한 발간 프로젝트가 추운 겨울이 되어서야 마무리되었다. 케언즈에서 졸업 후 쉬며 책 제작을 담당하겠다는 영상 회의를 했었는데 그 후 몇 달 동안 여행을 다니며 노트북으로 조금씩 작업을 하는 바람에 기존에 여름 출판 예정이었으나 반년 더 늦춰져 버렸다.

긴 시간동안 졸업생 아홉 명의 포트폴리오를 어떤 식으로 정리할지 많은 고민을 했다. 각자의 서사가 있고 색이 다르기에 느슨한 주제로 졸업 전시를 기획했던 것과 같이 몇 단계의 큰 틀 안에서 작품들을 보여주려했다. 창작물을 기록하는 아카이빙 작업에 있어 제일 중요한 것은 작가의 의도를 담아내는 것이기에 출판물로 보여지는 형식에 대한 많은 고민을 했다. 레이아웃과 한글 폰트, 영문 폰트, 내지 재질을 통해서 실험적이지만 단단한 알맹이를 가진 모두의 작업을 보여주고 싶었다.

학생 작업 하나하나 애정어린 글로 담아주신 우동선 선생님과 김태영 선생님, 2022년 한 해 동안 설계를 지도해주신 김수영, 김나리, 서승모, 캘빈츄아 선생님, 긴 시간 동안 기다려주시고 바쁜 와중에도 함께 책을 꾸리는 데 큰 도움을 주신 김태영 선생님과 김은지 리바조교님, 졸업심사 중에도 표지 디자인을 담당해주신 이정아 디자이너님, 인쇄 및 제책을 담당해주신 인타임, 마지막으로 멋진 작업을 담아내준 졸업생 강하림, 김보경, 김서진, 신지승, 이연우, 이은후, 전희수, 조은솔 모두에게 감사한 마음을 전한다.

이지윤
편집자, 한국예술종합학교 건축과 2023 졸업생

Acknowledgments

감사한 분들

고급스튜디오	김태영 선생님 김수영 선생님 서승모 선생님 김나리 선생님 캘빈츄아 선생님
기술스튜디오	정재학 선생님
전임 교수진	민현식 선생님 김봉렬 선생님 김종규 선생님 박선우 선생님 우동선 선생님 김태영 선생님 김병찬 선생님 이강민 선생님 지강일 선생님
초·중급스튜디오	이지은 선생님 최진석 선생님 최윤희 선생님 이도은 선생님 조경찬 선생님 이호선 선생님 이지영 선생님
이론 과목	김성완 선생님 최선영 선생님 이민아 선생님 이승환 선생님 최종훈 선생님 김민 선생님 박승진 선생님 박진택 선생님 민소정 선생님 박성중 선생님 김인철 선생님 김하나 선생님 황선우 선생님 이다미 선생님 이형진 선생님 심형섭 선생님

졸업도우미

강하림.	이혜정 이원준 이수 이명재 윤재원
김보경.	현은우 김지후 김동현 신재형 최카밀라 최시원
김서진.	허해인 양진모 윤감수 김연경 허진아 한지원
신지승.	이지민 강다현 김가은 김의하 이승헌 이지현 이한결 정선아 정선혜 정지우 조혁재
이연우.	유경림 김민형 고베니 김회연 유현욱 윤은 조은 이유신
이은후.	김민혁 김철휘 박유신 송윤선 표준우 박준성 최지아
이지윤.	김진솔 한지원 조윤서 정태유 조남우 김윤상
전희수.	김지우 박송이 이선재 이지수 박민서 이영서
조은솔.	안정원 이보영 박자영 김재준 김민경

Taeyoung Kim

Taeyoung Kim graduated from the Department of Architecture and Graduate School of Seoul National University. After working at Kiohun Architect & Associates, she served as a Senior Associate at Gensler London while pursuing her Design Ph.D. at the Bartlett School of Architecture. Currently, she is a Professor of Architecture at the Korea National University of Arts, teaching architectural design and collaborating on designs with U.TOPO Architecture Office. Her notable projects include the Eunhye Community Housing (Seoul Architecture Award, Korea Architecture Award 2018), Seongdong and Seongsu Book Maru projects (Public Building Award 2018, 2022), Square 181 (Seocho Architecture Award 2023), and the winning proposal for the 2022 Public Housing Design Competition, titled 'Half-level City, Corridor Community.'

Don-Son Woo

Don-Son Woo, a professor in the Department of Architecture at the Korea National University of Arts since September 2001, specializes in the history and theory of modern and contemporary architecture. He obtained his Dr. Eng. degree from the Graduate School at the University of Tokyo in March 1999. In 2008, he served as a visiting scholar at UC Berkeley, and in 2015, was a visiting foreign researcher at the Kyoto Institute of Technology (KIT). He was elected as the president of the Korean Society for Urban History in 2009 and as the president of the Korea Association for Architectural History in 2023 for the next two years. He has published 37 peer-reviewed papers, 47 conference papers, co-authored 24 books, translated 5 books, and conducted research on 14 oral historical monographs up to the present.

Seungmo Seo

Seungmo Seo was born in Kyoto in 1971, obtained his Master's degree of Fine art from Tokyo university of the Arts School of Architecture after his studies at Kyungwon University. He worked as a part-time lecturer at the college for two years, and then started working as an independent architecture in Seoul in 2004. He renamed Samuso Hyojadong in 2010 and has carried out a wide variety of projects including houses, hotels and offices.

김태영

김태영은 서울대학교 건축학과 및 대학원 졸업 후 기오헌건축사사무소를 거쳐 겐슬러 런던에서 시니어 어소시에이트로 근무하며 바틀렛 건축학교에서 디자인 박사과정을 밟았다. 현재 한국예술종합학교 건축과의 교수로 건축 설계를 가르치며 유토포건축사사무소와의 설계 협업을 병행하고 있다. 주요 작업으로 은혜공동체주택, 성동구 책마루 프로젝트, 방배동 스퀘어181 근린 생활시설, 2022년 대한민국 공공주택설계대전 당선안인 '반층도시, 복도공동체'가 있다.

우동선

우동선은 2001년 9월부터 현재까지 한국예술종합학교 미술원 건축과의 교수로서 근현대 건축의 역사와 이론을 담당하고 있다. 1999년 3월에 일본 도쿄대학(東京大學) 대학원 건축사연구실에서 박사학위를 취득하였고, 2008년의 유씨버클리(UC Berkeley) 방문학자와 2015년의 교토공예섬유대학(KIT) 국제방문연구원을 지냈다. 2019년에 도시사학회 회장을 역임하였고, 2023년 10월에는 2024년과 2025년의 한국건축역사학회 회장으로 선출되었다. 그는 현재까지 37편 심사논문, 47편의 발표논문, 24권의 공저, 5권의 역서, 14권의 채록연구를 발표하였다.

서승모

서승모는 1971년 일본 교토 출생으로, 경원대학교를 졸업하고 동경예술대학 건축학과에서 미술학 석사를 취득했다. 이후 2년간 동 학교 비상근 강사였으며, 2004년 서울에서 독립했다. 2010년 사무소명을 사무소효자동으로 개칭해 주거, 호텔, 업무시설 등 다방면으로 설계 영역을 넓혀가고 있다.

Naree Kim

Naree Kim studied architecture at Yonsei University in Korea and École Nationale d'Architecture Paris Val-de-Seine in France. After obtaining her degree of French Architect DPLG in 2006, she practiced at several architectural design firms. She has been working in France, Hong Kong, and Korea as a facade consultant for VS-A Group since 2010. In 2013, she co-founded VS-A Korea with Dutch architect Robert-Jan van Santen, and in 2018, she co-founded Ublo, a window company. She aims to practice architecture that considers humans and the earth together.

Sooyoung Kim

Sooyoung Kim founded Sumbi Architecture in 2010 and has been running it ever since. His aim is to watch with bated breath the process of how the universal language of architecture is influenced by the site and completed into a single building, as the Italian architect Antonio Monestiroli once said, "Our projects are completed in the specificity of the place and the universality of the plan. His office has built offices, factories, swimming pools, and kindergartens with clear systems and functions.

Calvin Chua

Calvin Chua is an architect and urban strategist working at the intersection of sustainable design, planning and advocacy. With over 10 years of experience in the built environment sector, Calvin has designed buildings and developed strategic masterplans for various cities and regional governments. Calvin currently leads Spatial Anatomy, a strategic design practice working on projects across scales, culture and geographies. Through a rigorous process of research and ground engagement, Spatial Anatomy has collaboratively designed buildings and regeneration masterplans across Asia alongside policy makers, grassroots and local governments.

김나리

김나리는 연세대학교 건축공학과와 프랑스 파리 발드센 국립건축학교에서 건축을 공부했다. 2006년 프랑스건축사(DPLG)를 취득하고 설계사무소에서 근무하다 2010년부터 브이에스에이 사무소 그룹 파사드 컨설턴트로 프랑스, 홍콩, 한국에서 일했다. 2013년 네덜란드 건축가 로버트-얀 반잔텐과 브이에스에이 코리아, 2018년 환기창을 개발하는 유블로를 공동 설립하여 운영 중이다. 인간과 지구를 함께 생각하는 건축을 실천하고자 한다.

김수영

김수영은 2010년에 숨비건축을 설립하여 현재까지 운영하고 있다. 그의 지향점은 '장소의 특수성과 계획의 보편성에서 우리의 프로젝트는 완성된다'고 한 이탈리아 건축가 안토니오 Antonio Monestiroli의 말처럼 건축의 보편적인 언어가 장소에 영향을 받으며 하나의 건축물로 완성되어가는 과정을 숨죽이고 응시하는 것이다. 그의 사무실에서는 명확한 시스템과 기능이 있는 오피스, 공장, 수영장, 유치원 등을 지어오고 있다.

캘빈츄아

건축가이자 도시 전략가로 지속 가능한 디자인, 계획 분야에서 일하고 있다. 건축 환경 분야에서 10년 이상의 경력을 쌓은 그는 다양한 도시와 지방 정부를 위해 건물을 설계하고 전략적 마스터플랜을 개발했다. 현재 그는 다양한 규모, 문화, 지역에 걸친 프로젝트를 수행하는 전략적 디자인 회사인 Spatial Anatomy를 이끌고 있다. Spatial Anatomy는 연구 및 현장 참여 프로세스를 통해 정책 입안자, 풀뿌리 및 지방 정부와 함께 아시아 전역의 건물 및 재생 마스터플랜을 공동으로 설계했다.

뒷줄: 이연우, 신지승, 김보경, 이지윤, 강하림,
앞줄: 조은솔, 김서진, 이은후, 전희수

Harim Kang

After graduating from Korea National University of Arts and Design, she worked as a researcher at the Passive Zero Energy Building Research Institute and provided training for ZEB experts for the Ministry of Land, Infrastructure, and Transport and the Korea Energy Agency. Her main works include the planning and production of the children's book «"0" House» and «Energy Independent Architecture with RNB Architects» to easily convey zero-energy architecture to future generations, and his awards include an encouragement in the bachelor's category of the 19th Korean Institute of Architects' Outstanding Graduation Thesis Exhibition on "The Impact of Florian Beigel's Landscape Architecture on the Paju Cultural Complex."

Bogyong Kim

Bogyong Kim studied at the design studio of Kim Sooyoung, Kim Jongkyu, and Seungmo Seo, and worked in the architecture and photography department for five years. In 2020, he won the grand prize at the Junglim Student Architecture Award 2021 Night Library. She interned at Sumbi Architects in the summer of 2021 and Cho Soeun Architects in the summer of 2022. Currently, she is gaining experience in architectural practice at Kim Hyo-young Architects.

Seojin Kim

Seojin Kim studied under the design studios of Jongkyu Kim, Kyungchan Choi, Seungmo Seo, and Taeyoung Kim + Nari Kim. In 2019, he won the Excellence Award at the 'Korean Apartment' research competition organized by Doosan Art Center, and in 2022, she won the fourth prize at the Architecture and Culture Award organized by the Shin Young Foundation. During her studies, she was an exchange student at Miroslav Sik's studio at the Academy of Visual Arts in Prague (AVU) in 2020, and after graduation, she worked as an overseas intern at Osamu Morishita Architect and Associates in Japan.

강하림

한국예술종합학교 건축과 졸업 후 패시브제로에너지건축연구소에서 연구원으로 근무하며 국토교통부 및 한국에너지공단의 ZEB 전문인력 양성교육 용역을 수행하고 있다. 주요 작업으로는 미래세대에 제로에너지건축을 쉽게 전달하기 위한 동화책 〈"0"이 되는 집〉, 〈RNB 건축사와 함께하는 에너지 자립 건축〉의 기획 및 제작이 있으며, 수상 이력으로는 "플로리안 베이겔의 랜드스케이프 건축이 파주 문화단지에 미친 영향"을 주제로 한 대한건축학회 19회 우수졸업논문전 학사부문 장려가 있다.

김보경

한국예술종합학교 미술원 건축과에 2018년도에 입학하여 김수영, 김종규, 서승모 설계스튜디오에서 공부하였고, 5년간 건축과 촬영부에서 활동하였다. 2020년도에 정림학생건축상 2021 밤의 도서관에서 대상을 수상한 바 있다. 2021년도 여름 숨비건축사사무소, 2022년도 여름 조소은건축사사무소에서 인턴쉽을 하였다. 현재 김효영건축사사무소에 입사하여 건축실무 경험을 쌓고 있다.

김서진

한국예술종합학교 미술원 건축과에 입학하여 김종규, 조경찬, 서승모, 김태영+김나리 설계스튜디오에서 수학했다. 2019년 두산아트센터에서 주관하는 '한국의 아파트' 리서치 공모전에서 우수상을 수상했으며, 2022년 신영문화재단에서 주최한 건축문화상에서 4등을 수상한 바 있다. 재학 당시 2020년도 체코 프라하예술학교(AVU)의 미로슬라프 식 (Miroslav Sik) 스튜디오에 교환학생 생활을 보냈으며, 졸업 이후 해외인턴으로 일본에 위치한 Osamu Morishita Architect and Associates에서 일하는 등 다양한 경험을 쌓고 있다.

Jiseung Shin

While studying architecture at Korea National University of Arts, Seoul, She was nominated for the RIBA Bronze Medal for my part 1 design work 'Where'. She held a solo exhibition, "The Part and The Whole," at the Corridor Gallery in the school, and exhibited my work in the Architecture and Photography Exhibition, the Advanced Studio Drawing Exhibition, and the Corridor Exhibition. She served as a student interviewee for KAAB and RIBA recertification. She has participated in the AA Visiting School Seoul twice. She received the 2nd Shin Young Foundation Award for Architecture and Culture for her graduation work, "The Mounds, Artifacts." She got The Korean Institute of Architects' 19th Outstanding Graduation Thesis Award for her RIBA part 2 thesis, "Archipelago-type Urban Plan for Architects Based on the Urban Theory Research of Pier Vittorio Aureli.". She completed RIBA part 2 and was awarded the Minister of Culture, Sports, and Tourism Award for the 2023 at the Korea National University of Arts upon graduation from the Department of Architecture. She has participated as an artist for the 2023 Seoul Urban Architecture Biennale Seoul Centennial Master Plan. She is currently practicing at architectural firm One O One Architects.

Yonu Lee

After graduating from the Korean National University of Arts and Sciences in Architecture, he is currently working at Simplex Architects, where he spent a working holiday at R.A.P in Rotterdam (2017), interned at ANU (2022), and served as a student journalist for the Korean Institute of Architects. Her participatory exhibitions include 'DESIGN IS MY RELIGION IS MY DESIGN' (ECI Culturefabriek, 2017), 'Seoul, Unfolded' (Korean Culture Center UK, 2020), 'Taller, Longer, Lighter' (Egan House Gallery, 2020), and 'AZIT: Please enter your search term' (Gallery Space 35, 2021) in collaboration with Dutch architectural firm R.A.P..

신지승

한국예술종합학교 건축과 재학 중 part 1 설계 『Where』로 RIBA Bronze Medal 후보에 올랐다. 미술원 복도 갤러리에서 개인전 《부분과 전체》를 열었고, 건축과 사진전, 건축과 고급 스튜디오 드로잉 전시, 복도 전시에 작업을 출품했다. KAAB와 RIBA 재인증 당시 학생 면담 대상자로 인증 갱신에 역할을 다했다. AA Visiting School Seoul 2회 참여했다. 졸업 작품 『The Mounds, Artifacts』로 제2회 신영문화재단 건축문화상 장려상, RIBA part2 논문 「피에르 비토리오 아우렐리의 이론을 기반으로 건축가의 군도archipelago형 도시계획안, 도시 이론 연구」로 대한건축학회 제19회 우수졸업논문전에서 장려상을 수상하였다. RIBA part 2를 수료하였으며, 한국예술종합학교 건축과 졸업과 동시에 2023 문화체육관광부 장관상을 수상했다. 2023 서울도시건축비엔날레 서울 100년 마스터플랜 참여 작가로 활동하였고, 현재는 건축사사무소 원오원 아키텍스에서 실무 수련을 하고 있다.

이연우

한국예술종합학교 건축과 졸업 후 현재 심플렉스건축사사무소에 재직 중으로 로테르담 R.A.P(2017)에서 워킹홀리데이와 ANU(2022)에서 인턴을 거쳤으며 한국건축가협회 학생기자로 활동하였다. 참여 전시로 네덜란드 건축사사무소 R.A.P와 협업한 'DESIGN IS MY RELIGION IS MY DESIGN' (ECI Cultuurfabriek, 2017), 'Seoul, Unfolded' (Korean Culture Centre UK, 2020), 'Taller, Longer, Lighter' (이건하우스 갤러리, 2020)와 'AZIT: 검색어를 입력하세요' (갤러리 공간35, 2021)가 있다.

Eunhoo Lee

Eunhoo Lee graduated from the Korea National University of Arts in Seoul from the Department of Architecture. She has planned and participated in exhibitions as an artist at 'Hyochang, Public Movement' in Seoul Metropolitan Government in 2019 and 'Resuscitation of Old Houses' in Mokpo in 2021, and received an encouragement award in the bachelor's category at the 19th Korean Institute of Architects Graduation Thesis Exhibition in 2023.

Jiyun Lee

Jiyun Lee graduated from Korea National University of Arts and is currently working at the architectural firm Samuso Hyoja. She is interested in the process of realizing architecture in response to external factors. In 2019, she completed a course in tropical climate architecture in Bali, Indonesia, and in 2021, she worked as an intern at architectural firm NOMAL. In 2023, she received the Bachelor's Prize at the Korean Institute of Architects Excellent Graduation Thesis Exhibition for her thesis comparing the dome construction methods of Orthodox churches in different regions in Korea.

Heesoo Jeon

Heesoo Jeon graduated from Korea National University of Arts, and her work focuses on eco-friendly infrastructure and urban regeneration. Her major off-campus works include being selected as the winner of the 'Urban Regeneration Resident Announcement Project for Yongsan Electronic Shopping District' in 2019 and exhibited at the Yongsan Y Valley ART CONTEST, and winning the grand prize at 'Seoul Through the Eyes of a Space Coordinator' in 2022, which was exhibited at the Total Art Museum. She is currently working as an intern at Irojae Architects.

Eunsol Jo

After graduating from Korea National University of Arts department of Architecture, Eunsol Jo worked at an architectural atelier in Geneva, Switzerland. She was selected and exhibited as a project for the Seoul Urban Architecture Biennale in 2023 and participated in the Art Basel Architecture Project in Switzerland. Eunsol Jo is currently involved in large and small projects in Switzerland.

이은후

한국예술종합학교 미술원 건축과를 졸업하고 현재 건축사사무소적재에서 근무하고 있다. 2019년 서울시청 '효창, 공공의 움직임', 2021년 목포 '폐가부활전, 낡은 집의 심폐소생술'에서 작가로서 전시를 기획하고 참여하였으며, 2023년 〈제19회 대한건축학회 우수졸업논문전〉 학사부문 장려상을 수상했다.

이지윤

한국예술종합학교 건축과를 졸업하고 건축사사무소 사무소효자에서 근무하고 있다. 외부요인에 조응하여 건축이 실현되는 과정에 관심이 있다. 2019년에는 인도네시아 발리에서 열대기후 건축과정을 수료하고, 2021년에는 건축사사무소 노말에서 인턴으로 근무했다. 2023년에는 대한건축학회 우수졸업논문전에서 지역에 따른 정교회 성당의 돔 축조방식을 비교한 논문으로 학사부문 장려상을 수상했다.

전희수

한국예술종합학교를 졸업했고, 친환경 인프라, 도시재생 등에 초점을 두어 작품 활동을 하였다. 주요 교외 작업으로는 2019년 '용산 전자상가 일대 도시재생 주민 공고사업'에서 당선작으로 선정되어 용산 y밸리 art contest에서 해당작품으로 전시한 바 있으며, 2022년 '스페이스 코디네이터의 눈으로 본 서울'에서 최우수상을 수상하여 토탈 미술관에서 전시한 바 있다. 현재 이로재 건축사사무소에서 인턴십으로 활동하고 있다.

조은솔

한국예술종합학교 건축과 졸업 후 스위스 제네바에 위치한 건축아틀리에에서 근무하고 있다. 2023년 서울도시건축비엔날레 프로젝트로 선정되어 전시하였으며 스위스 아트바젤 건축프로젝트에 참여하였다. 현재 스위스에 위치한 크고 작은 프로젝트에 참여하고있다.

UA Documents Vol.1

Department of Architecture,
School of Visual Arts
Korea National University of Arts
2022 Graduation Work

Portfolios
Harim Kang, Bogyong Kim, Seojin Kim,
Jiseung Shin, Yonu Lee, Eunhoo Lee
Jiyun Lee, Heesoo Jeon, Eunsol Jo

Contributors
Don-Son Woo, Taeyoung Kim,
Sooyoung Kim, Naree Kim,
Seungmo Seo, Calvin Chua, Jiyun Lee

Lead Editors
Taeyoung Kim, Don-Son Woo

Editorial Advisors
Jongkyu Kim, Byungchan Kim,
Kangmin Lee, Kangil Ji

Coordinators
Eunji Kim, Sanghyun Sung

Book Designer
Jiyun Lee

Cover Designer
Jiyun Lee

UA Documents Vol.1

한국예술종합학교 미술원 건축과
2022 졸업작품

포트폴리오
강하림, 김보경, 김서진,
신지승, 이연우, 이은후,
이지윤, 전희수, 조은솔

글.
우동선, 김태영, 김수영,
김나리, 서승모, 캘빈츄아,
이지윤

기획.
김태영, 우동선

자문.
김종규, 김병찬, 이강민,
지강일

조율.
김은지, 성상현

편집.
이지윤

표지디자인.
이지윤

UA Documents Vol.1
Back to the Earth

Department of Architecture,
School of Visual Arts
Korea National University of Arts
2022 Graduation Work

First published in December 12, 2024
Printed in Republic of Korea

Published by Zederolab.

Publisher
Kwiweon Chung

Design Manager
Minki Yoo (2mm)

15-2, Sajik-ro 8-gil, Jongno-gu,
Seoul, Republic of Korea
T.82 (0)2 2061 4146
zederolab2016@gmail.com

ISBN
979-11-987508-2-2 (03540)

© 2023 by Department of Architecture, School of Visual Arts, Korea National University of Arts and authors, all rights reserved including the right of reproduction in whole or in part in any form

UA Documents Vol.1
Back to the Earth

한국예술종합학교 미술원 건축과
2022 졸업작품

초판 1쇄 발행. 2024년 12월 12일
대한민국

펴낸곳.
제대로랩

펴낸이.
정귀원

디자인 감수.
유민기 (2mm)

서울시 종로구 사직로8길 15-2, 4층
T.82 (0)2 2061 4146
zederolab2016@gmail.com

ISBN
979-11-987508-2-2 (03540)

© 이 책의 모든 저작권은 저자에게 있으며 전부 또는 일부의 내용을 재사용하려면 저작권자의 동의를 받아야합니다.